Real-Time Monitoring of Cancer Cell Metabolism for Drug Testing

Hamed Alborzinia

Real-Time Monitoring of Cancer Cell Metabolism for Drug Testing

With a foreword by Prof. Dr. Stefan Wölfl

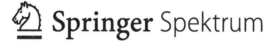 Springer Spektrum

Hamed Alborzinia
Heidelberg, Germany

Dissertation Heidelberg University, 2011

ISBN 978-3-658-10160-2   ISBN 978-3-658-10161-9 (eBook)
DOI 10.1007/978-3-658-10161-9

Library of Congress Control Number: 2015941017

Springer Spektrum
© Springer Fachmedien Wiesbaden 2015

Printed on acid-free paper

Springer Spektrum is a brand of Springer Fachmedien Wiesbaden
Springer Fachmedien Wiesbaden is part of Springer Science+Business Media
(www.springer.com)

To my parents, in love and appreciation!

Foreword by the Supervisor

Laboratory *in-vitro* models that mimic *in-vivo* conditions to follow the response of cells to changes in their micro-environment, including treatment with drugs and toxic compounds, is still a challenging and demanding aim for scientists. In particular for drug research it is very important to know all the effects that can be triggered in mammalian cells when exposed to a bioactive, potential drug or toxic compound. Such a system will be very useful in the field of medicinal chemistry as well as for pharmaceutical companies. Newly synthesized compounds designed for therapeutic applications may not only act on their designated targets, but rather interfere with various biochemical pathways, which could have negative and beneficial impact on their application. These effects have to be carefully monitored before drugs can be used in therapeutic applications in the clinic.

For this, all drug candidates and other chemical compounds are analyzed in great detail to understand their biological activities and modes of action. Nevertheless, many commonly used laboratory model systems are very limited. A major problem of most laboratory assays as well as animal studies is that the results are analyzed after a given treatment time in a so-called end-point assay. This means that after treatment is initiated the response to the treatment is only analyzed at predefined time points. Although, these assays provide very informative information, this is limited to these predefined time points. With the biosensor assay system implemented here for the analysis of drugs and other chemical compounds, the cellular response is monitored continuously and biochemical activity is recorded in real time.

Using a continuous flow system, important limitations of traditional tissue culture requiring high nutrient concentrations such as glucose or amino acids can be avoided and conditions closer to the physiological *in-vivo* nutrient condition can be used.

Thus, a system is needed, in which we will be able the monitor biological effects in real time and keep cells in a continuous-flow perfusion enabling to feed cells with physiological levels of nutrients.

In his research project, **Dr. Hamed Alborzinia** nicely presents the advantages of both time-resolved analysis of cells using online sensor measurement and continuous feeding in a flow-perfusion system. In his work he developed cell-based biosensor culture conditions with specific tissue properties that can be used for detailed analysis of cell metabolism. To learn different methods of sensor technology, he spent several months in the research lab of a small company, where he got introduced to biosensor chip technology and learned basic properties of electronic chip sensors. The initial experiments he performed there have been so successful that we obtained funding from the German Ministry for Research (BMBF) to set up this technology in our lab in Heidelberg. His work with this sensor technology led to a large number of important publications studying the activity of potential anticancer drugs, which could not have been described properly otherwise. Dr. Hamed Alborzinia established this technology as a highly efficient research platform and combined biosensor analysis with biochemical and other cell-based assays for more detailed molecular analysis to follow biological changes at key time points observed in the metabolic measurements. He also demonstrated that the biosensor system can be used to investigate the cellular response to growth factors, general toxic challenges, and to monitor the impact of important regulatory proteins such as SIRT3 and N-MYC on cancer cell metabolism. The impact of his thesis work is nicely visible in several peer-reviewed publications in medicinal chemistry and basic cancer biology and is very likely to attract significantly more attention in the future.

Prof. Dr. Stefan Wölfl

Acknowledgments

The aid and support of many people has made this possible after five long years of my Ph.D.

First, I would like to express my warmhearted gratitude to my advisor and supervisor Prof. Dr. Stefan Wölfl for his leadership, support, meticulous attention to details, his hard work, while yet allotting the necessary academic freedom and atmosphere to creatively work and develop my own ideas. Equally, my sincere thanks go to Prof. Dr. Jürgen Reichling, second advisor, for his kind support and helpful discussions and comments.

I am grateful to all members of the research network DFG FOR630 for fruitful collaboration, scientific support, and valuable discussions over the years.

My appreciative thanks are expressed to all of my laboratory colleagues for their ongoing support and professionality, in particular to: Dr. Catharina Scholl, Dr. Pavlo Holenya, Dr. Ngoc Van Bui, and Drs. Igor and Ana Kitanovic, as much as to Elke Lederer and Petra Fellhauer for technical and logistic assistance. Sincere thanks and gratitude are forwarded to Theodor C. H. Cole and Erika Siebert-Cole for their kind care and much appreciated logistic support during my studies.

To my friends over these years, I am particularly indebted: Dr. Steffen Walczak, Meike Büchler, Dr. Suzan Can, Dr. Christian Dransfeld, and Dr. Bettina Bradatsch.

Finally and most importantly I would like to thank my family: my wife, Marjan Shaikhkarami, for her unconditional support, and my beloved brothers, Dr. Reza Alborzinia and Dr. Hamid Alborzinia, who have been close, always, and who supported me in any possible way throughout my entire life. My utmost gratitude is owed to my parents, both of whom instilled in me the ever-important values of education, who always have believed in me, and who were always there for me, unconditionally! I am deeply indebted to them for their continued support and unwavering faith – this thesis is dedicated to them.

Hamed Alborzinia

Abbreviations

ADP	Adenosine diphosphate
ATCC	American Type Culture Collection
APAF	Apoptosis proteases activating factor
ATP	Adenosine-5′-triphosphate
CCCP	Carbonyl cyanide *m*-chlorophenyl hydrazone
Caspase	Cysteine/aspartate-specific proteases
*t*BHP	*tert*-Butyl hydroperoxide
CDDP	Cisplatinum or Cisplatin
cDNA	complementary DNA
CREB	cAMP response element repressor
DAVID	Database for Annotation, Visualization, and Integrated Discovery
DMEM	Dulbecco's modified Eagle's medium
DNA	Deoxyribonucleic acid
D-PBS	Dulbecco's phosphate buffered saline
EDTA	Ethylene diamine tetraacetic acid
ERK	Extracellular signal-regulated kinase
FACS	Fluorescence-activated cell sorting
FBS	Fetal bovine serum
FCS	Fetal calf serum
FDA	Food and Drug Administration (USA)
5-FU	5-Fluorouracil
GLUT	Glucose transporter
GSK-3ß	Glycogen synthase kinase-3beta
HEPES	4-(2-Hydroxyethyl)-1-piperazine ethane sulfonic acid
IC_{50}	Half maximal inhibitory concentration
IDES	Interdigitated electrode structure
ISFET	Ion-sensitive field effect transistor
MMS	Methyl methanesulfonate
mtDNA	mitochondrial DNA
NaF	Sodium fluoride
NAMI-A	New anti-tumor metastasis inhibitor-A
NIAID (NIH)	National Institute of Allergy and Infectious Diseases (National Institutes of Health)
p53	Tumor protein 53
PCR	Polymerase chain reaction
PI3K	Phosphatidylinositol 3-kinase
PKB/Akt	Protein kinase B/thymoma of the AKR mouse strain
PMSF	Phenylmethylsulfonyl fluoride
RM	Running medium
RNA	Ribonucleic acid
ROS	Reactive oxygen species
RT-PCR	Real-time PCR (or reverse transcriptase real-time PCR)
SDS	Sodium dodecyl sulfate
SGLT	Sodium glucose cotransporters
SIRT3	Sirtuin3
TrxR	Thioredoxin reductase

Contents

Summary

Analysis and identification of biological activities of drug and drug candidates is one of the most challenging tasks in modern drug research. To avoid unnecessary and costly tests in animal models it is very important that *in-vitro* test systems are available to provide detailed information regarding the biological activity and toxicity of potential drug candidates. Currently, most cell-based *in-vitro* bioanalytical methods used in pharmaceutical research are end-point measurements. This means that in each experimental assay only information for one particular time point is obtained, i.e.: *i*) cells are treated with compounds, *ii*) then at a preselected defined time point of interest the cells are fixed, lysed, or labeled, and *iii*) the resulting effects of the compound are monitored. By doing so, important information about the time dependence of biological activities are lost, or many repeated experiments have to be performed to cover all time points of interest.

To overcome this problem, novel biosensor chip analysis systems are enabling the continuous monitoring of cell metabolism and cell morphology in real-time, without any labeling or further disturbance of the system. The Bionas 2500 biosensor chip system used in this work allows the continuous monitoring of three important metabolic and morphological parameters: *i*) **oxygen consumption** using Clark-type electrodes, *ii*) **pH change** of the extracellular environment using ion-sensitive field effect transistors, and *iii*) the **impedance** between two interdigitated electrode structures to register the impedance under and across the cell layer on the chip surface. It also can be used with any adherent cell type, allowing further elucidation of specific drug properties.

In this thesis the biosensor chip was used to monitor the metabolic and morphological changes in five cancer cell lines in real-time in response to: (1) **cisplatin** (CDDP) treatment, one of the most widely used anticancer drugs; (2) overexpression of **sirtuin deacetylase SIRT3**, a key regulatory enzyme of cellular metabolism; and (3) a choice of several **organometallic compounds**, potential new anticancer drug candidates. To ensure that the observed parameters are of pharmacological relevance and not just an experimental artifact, further experimental analysis was performed to confirm the validity of the measured parameters. This included the role of the experimental conditions, like glucose concentration and uptake, but also detailed downstream analysis of molecular changes for the molecular interpretation of the observed results.

In the specific analysis of drug activity and molecular manipulation of the cells the following major results were obtained:

(1) All cell lines treated with **cisplatin** showed a first effect on respiration, which was followed by interference with glycolysis in four of the five cell lines, HT-29, HCT-116, HepG2, and MCF-7 but not in the cisplatin-resistant MDA-MB-231. Most strikingly, the cisplatin-sensitive cell lines start cell death within 10–11 h of treatment, indicating a clear timeline from first exposure to the drug, to cisplatin-induced lesions, and to cell fate decision. Further analysis at time points of most significant changes upon cisplatin treatment in the breast cancer cell line MCF-7 revealed important molecular changes underlying these activities. For this purpose, the phosphorylation of selected signal transduction mediators connected with cellular proliferation, as well as changes in gene expression, were analyzed in samples obtained directly from sensor chips at the time points when changes in glycolysis and impedance occurred. The reported online biosensor measurements reveal details in the timeline of metabolic responses to cisplatin treatment leading up to the onset of cell death.

(2) Overexpression of the metabolic regulator **SIRT3** led to an increase in cellular respiration of up to 35%. To ensure that this can indeed be attributed to the concentration of SIRT3 protein in the cells, the changes in protein levels were confirmed by Western blot directly from cells grown on the biosensor chips.

(3) The biological activity of potential **organometallic** drug candidates, containing the covalently bound (or chelated) metals, iron, rhodium, ruthenium, or gold, revealed not only antitumor activity but also unexpected striking biological activities. While most ruthenium complexes strongly reduced cell impedance but only slightly affected respiration and glycolysis, others immediately caused significant effects

on respiration or glycolysis. Cell-line and drug-specific responses were identified, confirming the versatility of these biosensor chip measurements.

In essence, this work provides *i*) real-time measurements of basic cancer cell metabolism of different cancer cell lines; *ii*) a detailed timeline of the metabolic response to cisplatin treatment and clear detection of the time span between start of cisplatin treatment and onset of cell death, which reflects the time required for the underlying molecular mechanisms of cell fate decision; *iii*) direct functional measurement of the biological activity of a key regulatory protein of cellular metabolism following the kinetic change in respiration upon SIRT3 overexpression; and *iv*) the time-resolved impact of several organometallic compounds on cell metabolism and cell morphology, including unexpected and not yet understood highly significant and specific effects on cell-cell interaction and adhesion.

Zusammenfassung

Die Analyse und Identifizierung der biologischen Wirksamkeit von zugelassenen und in der Testphase befindlichen Arzneistoffen ist eine der größten Herausforderung in der modernen Arzneimittelforschung. Die Verwendung von *in-vitro*-Systemen ermöglicht unnötige und aufwendige Untersuchungen in Tiermodellen zu vermeiden wobei detaillierte Informationen bezüglich der biologischen Wirksamkeit und der möglichen Toxizität potenzieller Arzneimittelkandidaten gewonnen werden können. Die meisten derzeit in der Pharmaforschung verwendeten zell-basierten bioanalytischen *in-vitro*-Methoden beruhen auf Endpunkt- Messungen. Das heißt, dass in jeder experimentellen Untersuchung nur Informationen für einen bestimmten Zeitpunkt gewonnen werden können: *i*) Zellen werden zunächst mit der Wirksubstanz behandelt, *ii*) daraufhin werden die Zellen zu einem vorbestimmten Zeitpunkt fixiert, lysiert oder markiert, und *iii*) die eingetreten Wirkung der Substanz festgestellt. Auf diese Weise gehen wertvolle Informationen zur zeitabhängigen Wirkung verloren, oder man muss viele solcher Messungen in Serie wiederholen um den gesamten Zeitraum zu erfassen.

Zur Vermeidung dieses Problems, wurden in letzter Zeit neuartige Biosensor-Chip Analysesysteme entwickelt, die eine kontinuierliche Messung von Stoffwechselvorgängen und Zellstrukturveränderungen in Echtzeit ermöglichen, ohne die Notwendigkeit von Markierungen oder andersartig störenden Eingriffen in das System. Das in dieser Arbeit verwendete Bionas 2500 Biosensor-Chip System ermöglicht kontinuierliche Messungen dreier wichtiger Stoffwechsel- und morphologischer Parameter: *i*) **Sauerstoffverbrauch** durch Clark-Elektroden, *ii*) **pH-Änderungen** des außerzellulären Milieus anhand von ionenempfindlichen Feldeffekt-Transistoren und *iii*) **Widerstand** zwischen zwei interdigitierten Elektroden, welches entsprechende Messergebnisse von unterhalb und entlang der auf der Chipoberfläche befindlichen Zellschicht liefert. Das System kann auch für jedwede Art von adhärenten Zelltypen verwendet werden, womit verschiedene spezifische Eigenschaften von Arzneistoffen untersucht werden können.

In dieser Doktorarbeit wurden mit dem Biosensor-Chip die Stoffwechsel- und morphologischen Veränderungen von fünf Krebszelllinien in Echtzeit unter den folgenden Bedingungen untersucht: (1) **Cisplatin** (CDDP)-Behandlung – dies ist eine der meistverwendeten Zytostatika in der Krebstherapie; (2) Überexpression von **Sirtuindeacetylase SIRT3** – einem Schlüsselenzym des Zellstoffwechsels; und (3) einer Auswahl verschiedener **organometallischer Verbindungen**, potenziellen neuen Zytostatika.

Um zu gewährleisten, dass die beobachteten Parameter auch pharmakologisch relevant sind, und nicht lediglich versuchsbedingte Artefakte darstellen, wurden weitere experimentelle Analysen durchgeführt, um die Aussagekraft der Messungen zu bestätigen. Dies beinhaltete die Überprüfung der Versuchsbedingungen, wie der Glukose-konzentration und -aufnahme, sowie eine ausführliche Downstream-Analyse von molekularen Veränderungen um eine molekulare Inter-pretation der erhaltenen Ergebnisse zu ermöglichen.

Die gezielte Analyse der Arzneistoffwirkungen und molekularen Zell-manipulation brachte folgende Hauptergebnisse:

(1) Alle Zelllinien die mit **Cisplatin** behandelt wurden reagierten zunächst mit einer Veränderung der Zellatmung, gefolgt von einem augenscheinlichen Eingriff in die Glykolyse bei vier der fünf Zellinien, HT-29, HCT-116, HepG2, und MCF-7, allerdings nicht bei der Cisplatin-resistenten MDA-MB-231. Überraschend reagierten die Cisplatin-empfindlichen Zellen durch einsetzenden Zelltod innerhalb von 10–11 Stunden nach Beginn der Behandlung, welches auf eine klare Zeit-abfolge vom ersten Arzneimittelkontakt, über das Auftreten von Cisplatin-bedingten Lesionen, bis zur Zellschicksalsentscheidung hinweist. Weitere Untersuchungen zum Zeitpunkt der signifikantesten Veränderungen nach Cisplatin-Behandlung der Brustkrebs-Zelllinie MCF-7 deutete auf dieser Wirkung zugrunde liegende wichtige molekulare Veränderungen hin. Um dies zu verstehen, wurde an Proben, die direkt von dem Sensorchip zum Zeitpunkt der Änderung von Glykolyse und Widerstand erhalten wurden, die Phosphorylierung einzelner Signal-Transduktionsmediatoren, die mit der Zellproliferation in Zusammenhang stehen untersucht, wie auch Veränderungen in der

Genexpression. Die Messungen ergaben Details im Zeitverlauf der Stoffwechselreaktion auf Cisplatin-Behandlung, die zum Einsetzen des Zelltods führen.

(2) Überexpression des Stoffwechselregulators **SIRT3** führt zu einer Erhöhung der Zellatmung um bis zu 35%. Um sicherzustellen, dass dies in der Tat auf die Konzentration des SIRT3-Proteins in den Zellen zurückzuführen ist, wurden die Veränderungen der Proteinmenge durch Western-Blot direkt anhand von Zellen des Biosensor-Chips bestätigt.

(3) Die biologische Wirksamkeit der **organometallischen** Arzneistoff-kandidaten, mit den kovalent gebundenen (oder chelierten) Metallen, Eisen, Rhodium, Ruthenium, oder Gold, zeigte nicht nur zytostatische Wirkung sondern auch unerwarteter maßen prägnante Einflüsse auf den Stoffwechsel. Während Ruthenium-Verbindungen den Widerstand stark verminderten und dabei die Zellatmung und Glykolyse nur schwach veränderten, bewirkten die anderen organometallischen Verbindungen sofort signifikante Veränderungen der Zellatmung und Glykolyse. Zelllinien- und arzneistoffspezifische Reaktionen konnten nachgewiesen werden, was die vielseitige Nützlichkeit dieser Biosensor-Chip-Messungen unterstreicht.

Diese Arbeit umfasst *i*) Echtzeit-Messungen des Krebszell-Stoffwechsels verschiedener Krebszelllinien; *ii*) eine ausführliche Zeitabfolge der Stoffwechselreaktionen nach Cisplatin-Behandlung und ein eindeutiger Nachweis der Zeitspanne zwischen Beginn der Cisplatin-Exposition und dem Einsetzen des Zelltods, was wiederum die Zeitspanne widerspiegelt, die benötigt wird damit der zugrunde liegende molekulare Mechanismus zur Zellschicksals-entscheidung führt; *iii*) direkt funktionelle Messungen der biologischen Wirksamkeit eines Schlüsselregulationsproteins des Zellstoffwechsels als Reaktion auf die kinetische Veränderung der Zellatmung durch SIRT3 Überexpression; und *iv*) in der Zeitabfolge dargestellte Wirkungen verschiedener organometallischer Verbindungen auf den Zellstoffwechsel und zell-morphologischen Veränderungen, zusammen mit unerwarteten und derzeit nicht erklärbaren höchst signifikanten und spezifischen Wirkungen auf Zell-Zell-Interaktionen und -Adhäsion.

1 Introduction

1.1 Drug Screening with New Perspectives

There is an urgent need for reliable **monitoring systems** in order to prove the **biological activity** and **safety** of the vast numbers of **novel pharmaceutical drugs**. The ultimate goal in drug development is the final approval for clinical application regarding drug safety and target efficacy (Marx and Sandig 2007). Any potential candidate compound must pass several phases of clinical testing (Fig. 1.1). A major approach has been to use laboratory animals or animal models, but the screening of thousands of novel compounds in animal models is impractical and unethical aside from being costly and moreover target identification of particular compounds in laboratory animal is cumbersome.

Most of the *in-vitro* assays that are performed to obtain information about the pharmaceutical compounds are so-called **endpoint assays**, meaning that the assay provides information only for a **particular chosen point of time**: cells are treated with compound and at the desired time point the cells are fixed/labeled and the effect of the compound is monitored. Of course one can apply several time points in order to obtain a **time–effect profile** of the compound, but this is laborious and time-consuming and associated with high expenses. The **dynamic nature of physiological responses** and cell signal transductions calls for a **continuous online monitoring** of cellular activities.

To these means, a novel **online cell biosensor chip system** has become available to monitor in **real time** the **physiological parameters of cells in response to drug treatment**. This allows **continuous and label-free** monitoring and analysis of the **time-resolved** impact of various know and unknown compounds on metabolic parameters.

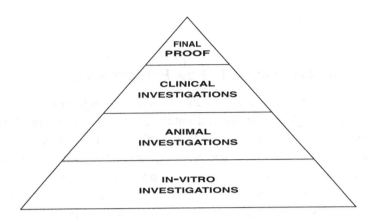

Fig. 1.1. Study phases for novel pharmaceutical compounds. *In-vitro* studies are followed by testing the compound on animals; treatment of a limited number of patients in clinical trials may lead to a first preliminary approval for clinical use; epidemiological studies on a representative population are then necessary for final approval (modified after: Marx and Sandig 2007)

1.2 Real-Time Monitoring of Living Cells Using a Novel Biosensor Chip

A new **biosensor system** (Bionas 2500; Fig. 1.2) has been designed to monitor cultured cells in real time, for several important physiological activities – with the ability of continuous measurement of the effects of drug-induced stress. A **silicon biosensor chip** serves as a platform for the growth of a monolayer of cells to be measured (Fig. 1.3) (Ehret et al. 2001).

In this thesis, three important metabolic parameters are surveyed in six different cancer cell lines in response to treatment with varying concentrations of cytotoxic drugs: **oxygen consumption** (respiration rate), **extracellular acidification** (glycolysis), and **cellular impedance** (reflecting morphological changes, cell–cell interactions, and membrane functionality) – features that mark the onset of drug-induced effects on metabolism and associated morphological changes.

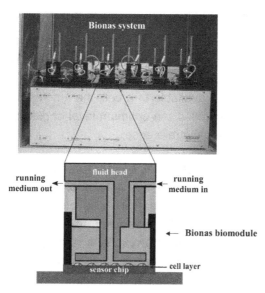

Fig. 1.2. Bionas 2500 system (Bionas biomodule) showing the flow system (*pink*) with inlet/outlet for running medium and the small incubation chamber; cell layer indicated on the sensor chip (Source: Bionas GmbH, modified, with permission; and from Alborzinia et al. 2011)

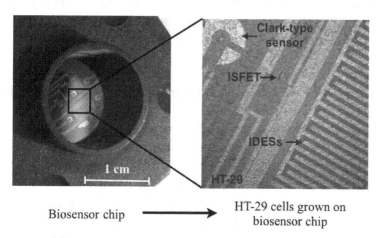

Fig. 1.3. Biosensor chip. Electrode sensors are visible and marked: ISFET (ion-sensitive field effect transistors): for pH; Clark-type sensor: for O_2; IDESs (interdigitated electrode structures): interdigitated electrodes for impedance (source: *left* Bionas GmbH, with permission; and from Alborzinia et al. 2011)

9

1.2.1 Oxygen Consumption

Variations in the cellular consumption of **oxygen** are generally indicative of enhanced or decreased mitochondrial activity – of **respiration**. As respiration is the major consumer of oxygen with direct dependence on the amount of **glucose** available to a cell, the amount of glucose in the medium should be controlled. An optimum level of glucose supply to the cells is necessary as a base reference.

Oxygen consumption can efficiently be measured by the so-called **Clark-type electrode**, which measures the amount of dissolved oxygen in the medium (Clark et al. 1953). There are also other ways for estimating mitochondrial activity, e.g., by measuring ATP – the essential final product of cellular respiration. However, as mentioned earlier, most of these procedures are end-point assays providing temporally restricted information.

Processes other than respiration are much less oxygen consuming and thus insignificantly contribute to this signal. Reactive oxygen species (ROS) are generated through mitochondrial activity and during energy production, but only a very small percentage of the total oxygen consumed by a cell is converted to ROS (Tarpey et al. 2004).

There are many processes that influence and control the metabolic balance in the cell. One key regulatory enzyme for the balance of cellular metabolism is the sirtuin 3 deacetylase SIRT3. It belongs to the sirtuin/mTOR (mammalian target of rapamycin) regulatory complex that balances the metabolic flux inside the cell in response to nutrient availability and growth signals. SIRT3 is located in the mitochondrial matrix where it regulates the activity of mitochondrial metabolic proteins by changing their acetylation level (Hirschey et al. 2010). The sirtuin/mTOR complex is crucial for the cellular metabolic balance and thus is intensively studied for its role in cellular aging processes and the linking of metabolic changes with aging. Nutrient availability and cellular metabolism are also believed to be a key factor in the sensitivity of tumor cells to drug treatment, directly linking mitochondrial activity with cellular drug response.

1.2.2 Extracellular Acidification

In **glycolysis** sugars are oxidized to produce ATP – when oxygen is limited, lactic acid is formed. Tumor cells are characterized by an increase of glycolysis and subsequent lactic acid fermentation (**Warburg effect**), and therefore glycolysis is an important parameter for studying the impact of cytotoxic drugs on cancer cells (Fig. 1.4). In these cells glycolysis levels can be efficiently estimated from the **extracellular pH** due to lactic acid formation measured via ISFET sensors (ion-sensitive field effect transistors).

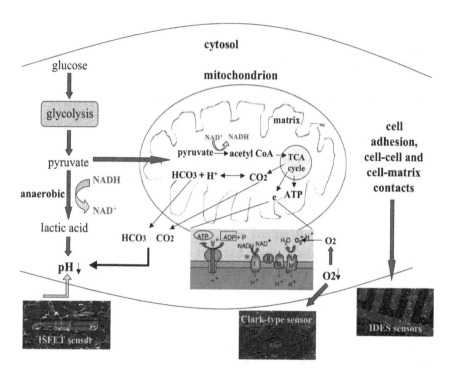

Fig. 1.4. Parameters used in the monitoring of cellular energy metabolism and adhesion via biosensor chips. Noninvasive measurements of extracellular pH (ISFET sensors), oxygen (Clark-type sensors), and impedance (IDES sensor) allow conclusions regarding cellular energy metabolism (i.e., glycolysis, respiration) and cell adhesion (source: *bottom three images* Bionas GmbH, with permission)

1.2.3 Cellular Impedance

Cell adhesion is a fundamental aspect of tissue formation, cell communication and proper functioning of cells. Measuring the **impedance** can provide information about **adhesion** and **morphological changes** regarding **membrane integrity** and **cellular vitality** and thus is useful for studying the effect of anticancer compounds on cancer cells. Powerful electrical techniques for measuring impedance are now allowing us to monitor minimal changes of cell morphology as well as cell-cell and cell-surface interactions (Giaever and Keese 1991).

IDES (interdigitated electrode structures) welded on biosensors can monitor the impedance changes caused by cells on the biosensor chip.

1.3 Anticancer Agents – Effects Measured in Real Time

1.3.1 Cisplatin – A Prominent Anticancer Drug

One of the most prominent compounds used in cancer chemotherapy is cisplatin (or cisplatinum, CDDP) a platinum-based inorganic compound (Fig. 1.5). Cisplatin has been used in cancer chemotherapy for more than 30 years against different types of human tumors, since its approval by the FDA in 1978 (Wiltshaw et al. 1976; Wittes et al. 1977). The substance was first described by Michel Peyrone in 1844 as "Peyrone's salt". Rosenberg et al. (1965) showed that growth of *Escherichia coli* was inhibited by an electric current delivered between two platinum electrodes. Later, the Rosenberg group was also able to prove that cisplatin can be used as a potent antitumor agent against solid tumors (Rosenberg et al. 1969). In the following years an enormous amount of research was devoted to an attempt of elucidating the mode of action of cisplatin. In principle, upon administration of cisplatin as a chemo-therapeutic agent it undergoes a process called *aquation*, particularly in this case meaning that one of the chloride ligands is displaced by water, because of the low intracellular chloride concentration. This leads to the formation of strong platinum covalent bonds with RNA, DNA (so-called DNA adducts), and proteins.

Fig. 1.5. Cisplatin

Despite extensive efforts, to date it has not been possible to fully elucidate the biological mode of action of cisplatin. DNA is the proven primary target of cisplatin, and as a result cisplatin adduct formation effects many DNA-dependent cellular functions, including inhibition of replication and transcription, cell cycle arrest, and DNA damage leading to cell death and apoptosis, but may also result in mutations (Chaney et al. 2005; Zdraveski et al. 2002; Wang and Lippard 2005; Vogelstein et al. 2000). Despite the severe renal toxicity of cisplatin, it is one of the three most common anticancer drugs. For testicular cancer the current success rate of treatment is ~90% in case of early diagnosis (Rabbani et al. 2001; Koberle et al. 1999; Shelley et al. 2002; Zamble et al. 1998). Cisplatin is also used against other tumors, such as ovarian, cervical, head and neck, esophageal, and non-small-cell lung cancer (Pegram et al. 2004), and shows promising effects in breast cancer treatment in combination with other drugs such as taxanes and trastuzumab (Koshy et al. 2010; Cassano et al. 1995). Several reports also have shown successful treatment of colon cancer with cisplatin in combination with other know compounds such as 5-FU (Gavin et al. 2008; Rebillard et al. 2007; Lacour et al. 2004; Jamieson and Lippard 1999).

Like cisplatin many other anticancer agents target DNA and chromosomes. However, chromosomal defects and dysfunctions of the DNA repair machinery and cell cycle checkpoints in tumor cells may render these compounds inactive. In these conditions, instead of preventing cell proliferation, these compounds may promote further DNA damage and genomic instability. Thus, instead of cancer cell elimination DNA targeting anticancer drugs could even promote further damage in DNA repair-deficient cells. This and other cancer-specific variations will increase their resistance to anticancer agents. In essence, the main goal of cancer treatment by anticancer drugs is the induction of apoptosis and cell death and inhibition of cellular proliferation of cancer cells without killing neighboring healthy cells. DNA damage caused by anticancer drugs normally activates the DNA damage response pathway through ATM and the response regulator p53. As a tumor suppressor, p53 can cause apoptosis upon severe DNA damage. This ideal response is blocked in some cancer cells carrying mutations in p53, which prevents induction of cell death upon DNA damage (Alberts et al. 2007).

Unlike other cells that can assume a final differentiated state without proliferation, tumor cells have two main options: one is survival and proliferation, and the other is cell death (key molecules involved in these pathways are presented in Fig. 1.6). Ras is an essential mediator transmitting signals from the periphery to the genome through MAP kinase, ERK, and PKB/Akt or GSK pathways. (Marks et al. 2009).

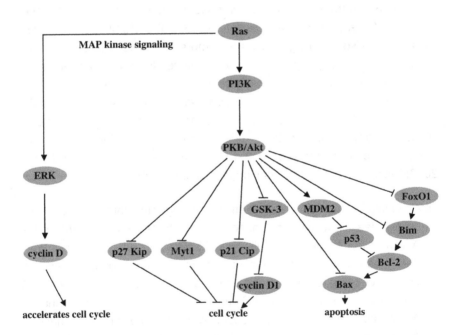

Fig. 1.6. Mitogenic signaling cascade associated with oncogenic mutations involved in cell cycle regulation and apoptosis (modified after Marks et al. 2009)

14

Proliferation and cell death (apoptosis) are controlled by both **extrinsic** and **intrinsic** signaling pathways. Growth factors are an example of extrinsic inducers of PI3K and Ras leading up to cell proliferation. An intrinsic pathway to apoptosis starts from DNA damage, which could result from cisplatin treatment. The resulting activation of p53 induces the transcription of Noxa and PUMA causing the release of cytochrome *c* and subsequent activation of caspase-9 leading up to apoptosis. Also effecting this pathway are such signaling molecules as Bcl2, Bax, Bak, and Bcl-XL which trigger or inhibit the release of cytochrome *c* by acting on mitochondrial membrane integrity (Marks et al. 2009) (Fig. 1.7).

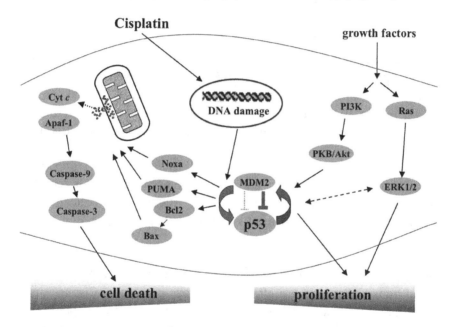

Fig. 1.7. Signaling pathway to cell proliferation and cell death. Growth factors induce PI3K and Ras leading up to cell proliferation. For this, p53 activity is also suppressed by MDM2. Cisplatin causes DNA damage resulting in activation of p53, which induces transcription of Noxa/PUMA; this triggers the release of cytochrome *c* and the subsequent activation of caspase-9 leading up to apoptosis. Bcl2, Bax, Bak, and Bcl-XL can trigger or inhibit the release of cytochrome *c* by acting on mitochondrial membrane integrity (modified after Marks et al. 2009).

1.3.1.1 Effect on Mitochondrial Activity

Mitochondria are the main target of cisplatin-mediated DNA damage (Garrido et al. 2008; Cullen et al. 2007) and important targets of apoptosis-inducing factors leading to cytochrome *c* release causing activation of caspase-9. In addition, cisplatin appears to accumulate in mitochondria, binding to mitochondrial DNA (mtDNA), and thus causing mitochondrial damage and ROS formation. The question thus arose as to whether cisplatin directly affects mitochondrial respiration and the formation of ROS. This interrelationship was investigated using a mitochondrial activity assay with isolated mouse liver mitochondria.

1.3.1.2 Phosphorylation of Akt1/ERK

The major regulatory pathways for cellular survival and proliferation are the PKB/Akt pathway and the Ras-MAPK pathway, which in response to growth factors and nutrient availability regulate cellular survival. These pathways are further connected to signals controlling cellular differentiation, e.g., wnt/gsk-3β/β-catenin and stress response PKA/p-CREB.

PKB/Akt (serine/threonine-specific protein kinase B) is an important signal transduction mediator and plays an essential role in cellular proliferation and apoptosis (Datta et al. 1999) being activated by phosphatidylinositol 3-kinases (PI3-kinases or PI3Ks) (Marks et al. 2009) (Fig. 1.6).

Members of the MAPK protein family, including **ERK1/2**, are involved in signal transduction and are affected by stress conditions. Cisplatin has been associated with the specific activation of signaling pathways mediating DNA damage response and cellular proliferation – including **p53**, **ERK1/2**, and **Akt1** (Kim et al. 2008; Datta et al. 1999; Panka et al. 2008). ERK1 contributes to apoptosis by activating the tumor suppressor p53 (Wang et al. 2000; Kim et al. 2005). Cisplatin can trigger the ERK pathway in tumor cells and through p53 activation can initiate cell apoptosis and cell cycle arrest (Kim et al. 2008).

1.3.1.3 Gene Expression

Online monitoring of cellular activities in response to drug treatment allows us to determine in more detail the actual time points of metabolic changes. Besides changes in signal transduction, the most significant change in cellular activity required change in gene expression. In this regard DNA microarrays have served to analyze the according gene expression in a cisplatin-sensitive breast cancer cell line (MCF-7) treated with cisplatin at specific time points identified as relevant by online measurements. To these means, treatment was performed in the biosensor chip system and cells were subsequently lysed so that RNA could be collected directly from the biosensor chip.

1.3.2 Organometallic Compounds

Cisplatin success in cancer chemotherapy for more than 30 years against different types of human tumors made medicinal chemist interested if other more complex compounds like organometallic compounds can be used in cancer treatment in more target specific than cisplatin. Organometallic compounds contain at least one covalent metal-carbon bond. Metals have variously been used for pharmaceutical treatment and metal complexes have shown interesting preclinical and clinical results as antitumor drugs. The use of metals or metal-containing compounds against ulcerous conditions but also in therapeutic treatment of cancer and leukemia dates back to the sixteenth century (Lange et al. 2008). The more advanced metal-based NAMI-A (ruthenium-containing) is very promising in clinical applications (Hartinger et al. 2006) (Fig. 1.8). Also chelated gold complexes inhibit tumor cell growth and many of these compounds strongly and specifically inhibit thioredoxin reductase (TrxR) (Ott 2009). However, the sensitivity of organometallics to air and water reduce their applicability for chemotherapy and heavy metals also often bear the risk of toxic side effects.

In this study we have tested several antitumor organometallic compounds – containing iron, rhodium, ruthenium, or gold – on several tumor cell lines.

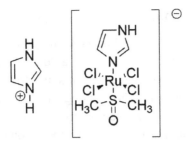

Fig. 1.8. NAMI-A

1.4 Aim of this Study

The biological activities of drugs are associated with complex physiological process – usually there is not only one single response leading up to one single effect, but rather there is a combination of drug responses involved. An important prerequisite for understanding the complexity of the cellular response to drug treatment is the **recording of the cellular response over time**, reflecting specific drug effects in a time-dependent manner.

However, not all cellular parameters can be analyzed in a time-dependent manner. With the recently available technique of **biosensor chip-based real-time monitoring** of cellular **metabolism** and **morphology** it is now possible to continuously monitor the cellular response to drug treatment.

The aim of this work has been to demonstrate the general applicability of this new biosensor system for analyzing the biological effects of anticancer drugs. As a first step, the well-known anticancer drug **cisplatin** has been studied to understand the basic readouts of the system. The results allowed the optimization of the experimental conditions and an understanding of the role of basic metabolism in drug response. Then the new established system was used for characterizing several new **organometallic** compounds and studying the effect of protein overexpression of a key regulatory enzyme of cellular metabolism, **SIRT3**.

The results obtained with the cell biosensor chip needed to be combined with further molecular analysis of the **cellular mechanisms** in order to better understand the significance of the new measurable parameters for drug activity. With this aim, cells were treated with cisplatin and harvested at time points of most significant metabolic changes (**respiration**, **glycolysis**, and **impedance**) which then were screened for changes in **signal transduction** associated with cellular proliferation and apoptosis, as well as for **gene expression** reflecting significant changes in cellular properties.

Core objectives:

- real-time analysis of cellular metabolism and morphological changes in response to drug treatment (cisplatin, several organometallics) using cell biosensor chips.
- optimization of the cell biosensor chip assay for screening anticancer drug activity.
- specificity of the signals in respect to metabolic activity and nutrient availability.
- detailed analysis of potential cellular mechanisms at crucial time points:
    - mitochondrial activity
    - specific activity of key signal transduction mediators
    - gene expression analysis

This thesis is the first comprehensive real-time study of the cellular response to the well-established anticancer drug **cisplatin** – and a reference for the use of the new online monitoring biosensor chip system in drug research.

2 Materials

2.1 Instruments

Bionas 2500	Bionas GmbH, Rostock, Germany
Cell culture incubator	Thermo Scientific, Massachusetts, USA
Laminar air flow	Integra bioscience, Zizers, Switzerland
Centrifuge 5702	Eppendorf, Hamburg, Germany
Stereomicroscope	Zeiss, Jena, Germany
Laboratory scale	Sartorious, Gottingen, Germany
Tecan Safire^2 microplate reader	Tecan, Maennedorf, Switzerland
Electroporation apparatus	Amaxa/Lonza, Köln, Germany
Gel documentation system	Peqlab, Erlangen, Germany
PCR mastercycler	Eppendorf, Hamburg, Germany
LightCycler® 480	Roche Appl. Science, Penzberg, Germany
Arraymate reader	Alere Techn. GmbH, Jena, Germany
FACSCalibur	Becton Dickinson, New Jersey, USA
pH-meter	Hanna Instruments, Rhode Island, USA
Pipettes P2, P20, P100, P1000	Eppendorf, Hamburg, Germany
Multistep-Pipette	Eppendorf, Hamburg, Germany
Light microscope, Motic AE31	Motic, Xiamen, China
Photometer	Thermo Spectronic, Madison, WI, USA
Vortex Genie 2	Bender & Hobein, Bruchsal, Germany
NanoDrop 2000c	Thermo Scientific, Massachusetts, USA
Water bath	Neolab, Heidelberg, Germany
Water purification system (EASYpureII)	Barnstead Intl., Dubuque, IA, USA
Autoclave VX-95	Systec, Wettenberg, Germany

2.2 Laboratory Materials

Cell culture petri dishes	Greiner bio-one, Kremsmuenster, Austria
Cell culture flask	Greiner bio-one, Kremsmuenster, Austria
Bionas metabolic chip	Bionas GmbH, Rostock, Germany
Eppendorf test tubes (1.5/2.0 mL)	Eppendorf, Hamburg, Germany
Polypropylene tubes (15 and 50 mL)	Greiner bio-one, Kremsmuenster, Austria
Pipettes (serological, sterile)	Greiner bio-one, Kremsmuenster, Austria
Flow cytometry tube	Becton Dickinson, New Jersey, USA
Array tube™	Clondiag/Alere GmbH, Jena, Germany
Affymetrix GeneChip® array U133 2.0plus	Affymetrix, Santa Clara, USA
Oxoplate (96 well)	PreSens/BPSA,Wash. DC, USA
Microtiter plates (96-, 24-, 6-well)	Greiner bio-one, Kremsmuenster, Austria

2.3 Chemicals

Cisplatin	Sigma-Aldrich, Missouri, USA
Ruthenium complexes	Prof.Dr. Schatzschneider (Univ. Würzburg)
Rhodium complexes	Prof.Dr. Sheldrick (Univ. Bochum)
[Salophene]iron complex	Prof.Dr. Gust (Univ. Innsbruck)
Benzimidazol-2-ylidenes gold complexes	Prof.Dr. Ott (Univ. Braunschweig)
Phloretin	Sigma-Aldrich, Missouri, USA
Phloridzin (Phlorizin)	Sigma-Aldrich, Missouri, USA
Penicillin/Streptomycin	Gibco, Carlsbad, USA
Fetal calf serum (FCS)	PAA, Pasching, Austria
Fetal bovine serum (FBS), fatty acid-free	PAA, Pasching, Austria
Agarose	Roche, Penzberg, Germany
Triton X-100	Sigma-Aldrich, Missouri, USA
Phenol Red	Sigma-Aldrich, Missouri, USA
HEPES	Sigma-Aldrich, Missouri, USA
Ethanol	Sigma-Aldrich, Missouri, USA

Nanofectin	PAA, Pasching, Austria
Glucose	Sigma-Aldrich, Missouri, USA
Sucrose	Sigma-Aldrich, Missouri, USA
Mannitol	Sigma-Aldrich, Missouri, USA
Malate	Sigma-Aldrich, Missouri, USA
Na_2SO_3	Sigma-Aldrich, Missouri, USA
KCl	Sigma-Aldrich, Missouri, USA
$MgCl_2$	Sigma-Aldrich, Missouri, USA
KH_2PO_4	Sigma-Aldrich, Missouri, USA
EDTA	Sigma-Aldrich, Missouri, USA
Tris	Sigma-Aldrich, Missouri, USA
Pyruvate	Sigma-Aldrich, Missouri, USA
ADP	Sigma-Aldrich, Missouri, USA
ATP	Sigma-Aldrich, Missouri, USA
Rotenone	Sigma-Aldrich, Missouri, USA
CCCP (carbonyl cyanide 3-chlorophenylhydrazone)	Sigma-Aldrich, Missouri, USA
Sodium dodecyl sulfate (SDS)	Merck, Darmstadt, Germany
ß-Mercaptoethanol	Merck, Darmstadt, Germany
Urea	Sigma-Aldrich, Missouri, USA
Aprotinin	AppliChem GmbH, Darmstadt, Germany
Sodium orthovanadate	AppliChem GmbH, Darmstadt, Germany
Sodium fluoride (NaF)	AppliChem GmbH, Darmstadt, Germany
Phenylmethanesulfonylfluoride (PMSF)	AppliChem GmbH, Darmstadt, Germany
Dihydroethidium	Sigma-Aldrich, Missouri, USA

2.4 Kit Systems

RNeasy Mini	Qiagen, Hilden, Germany
RevertAid™ Premium First Strand cDNA Synthesis	Fermentas/Thermo Fisher, USA
Western lightning chemoluminescence reagent	PerkinElmer LifeScience, USA
QIAprep®Miniprep Kit	Qiagen, Hilden, Germany
QIAquick PCR Purification Kit	Qiagen, Hilden, Germany

2.5 Primers and Oligonucleotides

List of forward and reverse primers used to perform quantitative real-time PCR

ATF3-5s	5′ – CATCCAGAACAAGCACCTC – 3′
ATF3-3as	5′ – GCATTCACACTTTCCAGC – 3′
CDK6-5s	5′ – GATGTGTGCACAGTGTCACGAAC – 3′
CDK6-3as	5′ – GTGGTTTTAGATCGCGATGCAC – 3′
E2F3-5s	5′ – GTTCATTCAGCTCCTGAGCCAG – 3′
E2F3-3as	5′ – CACTTCTTTTGACAGGCCTTGAC – 3′
SESN1-5s	5′ – CTGAAGAGCATCCAGGAAC – 3′
SESN1-3as	5′ – GCAGTAGATAGTGCTGAG – 3′
MALAT-5s	5′ – GGATCCTAGACCAGCATGC – 3′
MALAT-3as	5′ – GGTTACCATAAGTAAGTTCCAG – 3′
GADD45A-5s	5′ – CGATAACGTGGTGTTGTGC – 3′
GADD45A-3as	5′ – GAATGTGGATTCGTCACC – 3′
PMAIP1-5s	5′ – ATGCCTGGGAAGAAGG– 3′
PMAIP1-3as	5′ – CAGGTTCCTGAGCAGAAG– 3′
Human β-actin-5s	5′ – CTGACTACCTCATGAAGATCCTC– 3′
Human β-actin-3as	5′ – CTGCTGAAGAAGCACATCGATTC– 3′

2.6 Biological Software

Bionas analysis software

dChip software (Li and Wong, Harvard University)

DAVID Bioinformatics Resources 6.7 (Database for Annotation, Visualization, and Integrated Discovery) of NIAID (NIH) (http://david.abcc.ncifcrf.gov/home.jsp)

SerialCloner 1–3 (http://serialbasics.free.fr/Serial_Cloner.html)

2.7 Mammalian Cell Lines

HT-29 (human colon carcinoma)	ATCC, Manassas, USA
HCT116 (human colon carcinoma)	ATCC, Manassas, USA
HepG2 (human liver hepatocellular carcinoma)	ATCC, Manassas, USA
MDA-MB-231 (human breast adenocarcinoma)	ATCC, Manassas, USA
MCF-7 (human breast adenocarcinoma)	ATCC, Manassas, USA
HeLa (human cervix adenocarcinoma)	ATCC, Manassas, USA

2.8 Vectors

pcDNA3.1-hSIRT3$_{WT}$-Flag

pcDNA3.1-hSIRT3$_{H248Y}$-Flag

pcDNA3.1-hSIRT3$_{R80W}$-Flag

2.9 Media/Buffers/Solutions

Running medium (RM): DMEM without carbonate buffer and glucose (PAN Biotech GmbH, Aidenbach, Germany)

- 1 g/L glucose
- 1 mM Hepes
- 2 mM L-glutamin
- 100 units/mL penicillin
- 100 μg/mL streptomycin
- 15 μg/L phenol red
- 0.1% FCS

DMEM (Dulbecco's Modification of Eagles Medium) (100 units/mL penicillin, 100 μg/mL streptomycin, 10% FCS)	Gibco Invitrogen, Carlsbad, USA
Leibovitz's L-15 Medium (penicillin/streptomycin (100 units/mL penicillin, 100 μg/mL streptomycin 10% FCS, 1% nonessential amino acids)	Gibco Invitrogen, Carlsbad, USA
TrypLE ™ Express	Gibco Invitrogen, Carlsbad, USA
D-PBS (Dulbecco's phosphate buffered saline)	Gibco Invitrogen, Carlsbad, USA

Respiration buffer (pH 7.4)   25 mM sucrose
        100 mM KCl
        75 mM mannitol
        5 mM $MgCl_2$
        10 mM KH_2PO_4
        0.5 mM EDTA
        10 mM Tris
        0.1% fatty acid-free BSA
        10 mM pyruvate
        2 mM malate
        2 mM ADP
        0.5 mM ATP

Oxygen-free water (1% Na_2SO_3)

FACS buffer (D-PBS and 1% BSA)

Lysis buffer in PBS (pH 7.2–7.4) (6 M urea, 1 mM EDTA, 5 mM NaF, 0.5% Triton X-100)

2.10 Antibodies for Microarrays

phospho-GSK-3ß(S9)	R&D Systems, Minneapolis, USA
phospho-Akt1(S473)	R&D Systems, Minneapolis, USA
phospho-ERK1(T202/Y204)	R&D Systems, Minneapolis, USA
phospho-ERK2(T185/Y187)	R&D Systems, Minneapolis, USA
phospho-CREB(S133)	R&D Systems, Minneapolis, USA
Human paxillin	R&D Systems, Minneapolis, USA

3 Methods

3.1 Cell Culture

The applied cell lines were HT-29, HCT116, HepG2, MDA-MB-231, MCF-7 and HeLa (all from ATCC); all were cultured in Dulbecco's modified eagle medium (DMEM) (PAA, Pasching, Austria), supplemented with 10% FBS (PAA, Pasching, Austria), 1% penicillin/streptomycin (Gibco Invitrogen, Carlsbad, USA) within a CO_2 incubator at 37°C with 5% CO_2. Subculturing or splitting was performed regularly every 2–3 days. For the MDA-MB-231 cell line we used Leibovitz's L-15 Medium (Gibco Invitrogen, Carlsbad, USA) and nonessential amino acids (Gibco Invitrogen, Carlsbad, USA).

3.2 Real-Time Monitoring of Cellular Metabolism in Living Cells

3.2.1 Preparation of Running Medium

Ultrapure water obtained from water purification system (EASYpureII, Barnstead Intl. Dubuque, IA, USA) was autoclaved. DMEM powder medium (PAN Biotech GmbH, Aidenbach, Germany) was added according to manufacturer's procedure and in addition glucose (1 g/L), Hepes (1 mM), and 15 μg/L phenol red were added. After stirring the solution was adjusted to pH 7.4. Then the solution was filtered using Millipore filter. After addition of penicillin/streptomycin and 0.1% FCS, the medium is ready to use.

3.2.2 Bionas System Setup for the Experiment

Before starting the experiment, the Bionas reference electrodes must be checked to guarantee that sufficient electrolyte is available. Then the Bionas controlling software is started and the following steps selected for the experiments:

(*i*) disinfection: disinfecting the system before operation. For this we run 70% ethanol for 2 min with the flow rate of 100% (highest pump speed) and then 10 min 70% ethanol with flow rate of 4%. Then we rinse the system for 4 min with PBS with a flow rate of 25% and at the end we rinse the system with running medium (RM) for 4 min with a flow rate of 25%.

(*ii*) measurement: inserting the biosensor chip into the system and starting the experiment. During the measurement the flow rate is always set to 1%.

(*iii*) cleaning: for thorough cleaning of the system after each experiment we use 70% ethanol for 30 min with a flow rate of 4% and at the end we wash the system with ultrapure water for 4 min at 100% flow rate.

(*iv*) finish: in this step the pump is working for 2 min with the highest speed to remove remaining liquid from the system.

3.2.3 Preparation and Cultivation of Cells on the Chip

All cell lines were seeded at a density of 2×10^5 on each chip in 450 μL DMEM except MCF-7 with 1.5×10^5 per chip. After seeding, cells were grown on the chip for 20–24 h in standard cell culture conditions to approx. 80–90% confluence on the chip surface at the time of measurement. At the time of inserting the chip into the system an amount of 200 μL of medium were removed from the chip and cellular density checked under the stereomicroscope. Biosensor chips were placed into the Bionas system in the proper position and the experiment was started.

3.2.4 Real-Time Measurement of Cellular Metabolism

Changes in cellular metabolism and morphology were analyzed using a Bionas 2500 sensor chip system (Bionas, Rostock, Germany). The sensor chips (SC1000) allow the continuous measurement of three important parameters of cellular metabolism: (*i*) oxygen consumption using Clark-type electrodes, (*ii*) change in the pH of the extracellular environment using ion-sensitive field effect transistors, (*iii*) and the impedance between two interdigitated electrode structures to register the impedance under and across the cell layer on the chip surface.

As mentioned before, cells were seeded on the sensor chip in DMEM with penicillin/streptomycin and 10% (v/v) FCS (PAA, Pasching, Austria) and incubated in a standard tissue culture incubator at 37°C, 5% CO_2, and 95% humidity for 24 h until 80–90% confluence was reached. Sensor chips with cells were then transferred to the Bionas 2500 analyzer in which medium is continuously exchanged in 8-min cycles (4 min exchange of medium and 4 min without flow) during which the parameters were measured. The running medium (RM) used during analysis was DMEM without carbonate buffer (PAN Cat.Nr. P03-0010) and only weakly buffered with 1 mM Hepes and reduced FCS (0.1%) and low glucose (1 g/L). For drug activity testing, the four following steps were included: (a) 5 h of equilibration with only running medium with 4-min stop/flow incubation intervals, (b) drug incubation with compounds freshly dissolved in medium at indicated concentrations also with 4-min stop/flow, and (c) a drug-free step in which cells are again fed with RM without compound; (d) at the end of each experiment, cell layers are removed by adding 0.2% Triton X-100 to obtain basic signals without living cells on the sensor surface as negative control.

3.2.5 SIRT3 Overexpression – Experimental Setup and Real-Time Measurement

HeLa cells (ATCC, Manassas, USA) were transfected with either Nanofectin (PAA, Pasching, Austria) or Amaxa electroporation (Amaxa/Lonza, Köln, Germany) with nucleofection solution R (Lonza, Köln, Germany) –giving best transfection efficiencies out of all tested methods (nucleofection and Nanofectin versus lipofection and calcium–phosphate transfection).

Nanofection was performed according to the manufacturers' instructions, with the following modifications for use in the biosensor chips: HeLa cells were transfected in six well plates at 70–80% confluence with 200 μL transfection reagent (150 mM NaCl) per well to which 1.5 μg plasmid DNA and 9.6 μL nanofectin were added, followed by medium exchange with DMEM supplemented with 5% FBS after 5 h of transfection. After transfection (20 h) cells were trypsinized and seeded at a density of 3×10^5 cells on poly-D-lysine (Sigma, Deisenhofen, Germany) coated biosensor chips* in 450 μL DMEM with 5% FBS for 4 h prior to Bionas analysis. Amaxa nucleofections were performed according to the manufacturers' instructions using 10^6 cells per transfection out of which one third (3×10^5) were transferred onto poly-D-lysine coated biosensor chips in 450 μL Optimem (Gibco Invitrogen, Carlsbad, CA, USA) supplemented with 5% FBS and 1% penicillin/streptomycin for 4 h before Bionas kinetic measurements were started (Dransfeld et al. 2010).

Cells were seeded and grown to approx. 80–90% confluence on the coated chips. For analysis in the Bionas 2500 system the running medium was supplemented with 2.5% FCS and was continuously exchanged in 8-min cycles (4 min exchange of medium and 4 min without flow) during which the parameters were measured for 20–24 h after transfection for direct comparison of respiration and glycolysis or 4 h after transfection for kinetic analyses (Dransfeld et al. 2010).

* Coating: the biosensor chips (SC1000) were washed, disinfected, and coated with 100 μL poly-D-lysine (Sigma, Deisenhofen, Germany) for 5 min, followed by three washes with autoclaved ddH$_2$O and subsequent drying for 2 h.

3.3 Mitochondrial Oxygen Consumption

The measurement was performed using OxoPlate® (PreSens/BPSA, Wash. DC, USA), 96-well plates which contain an immobilized oxygen sensor at the bottom of each well. Fluorescence is measured in dual mode, excitation 540 nm, and emission 650 nm, with reference emission 590 nm. Their signal ratio 650/590 nm corresponds to the oxygen partial pressure. The calibration of the fluorescence reader is performed using a two-point calibration with oxygen-free water (1% Na$_2$SO$_3$) and air-saturated water with oxygen partial pressure corresponding to 0% and 100%, respectively. An amount of 18 μg of freshly isolated mitochondria were suspended in 100 μL of Respiration Buffer (25 mM sucrose, 100 mM KCl, 75 mM mannitol, 5 mM MgCl$_2$, 10 mM KH$_2$PO$_4$, 0.5 mM EDTA, 10 mM Tris, 0.1% fatty acid-free BSA, pH 7.4) containing 10 mM pyruvate, 2 mM malate, 2 mM ADP, and 0.5 mM ATP to activate oxidative phosphorylation. Cisplatin was diluted so that the appropriate concentration in the wells was established after adding the mitochondria.

Fluorescence was measured continuously for 400 min with kinetic intervals of 5 min by a Tecan Safire[2] (Tecan, Maennedorf, Switzerland) microplate reader at 37°C. During the measurements the plates were sealed with a breathable membrane (Diversified Biotech, Boston, MA). Additional controls were 5 μM rotenone (Sigma-Aldrich, Missouri, USA) as inhibitor of respiratory chain complex I and 1 μM CCCP (carbonyl cyanide 3-chlorophenylhydrazone, Sigma-Aldrich, Missouri, USA) as uncoupling agent, the latter of which is capable of increasing electron flow through the respiratory chain thereby increasing the oxygen consumption.

3.4 Antibodies and Recombinant Protein Standards for Protein Microarrays

Antibody microarrays (Wölfl et al. 2005) based on ArrayTube[TM] platform (Clondiag/Alere Techn. GmbH, Jena, Germany) utilize isotype-specific capture antibodies and biotinylated phospho-specific detection antibodies against five human/mouse/rat proteins of interest: phospho-GSK-3ß(S9), phospho-Akt1(S473), phospho-ERK1(T202/Y204), phospho-ERK2(T185/Y187), and phospho-CREB(S133). The microarrays used also contain an antibody against human paxillin. Microarray calibration was performed with recombinant protein standards. The antibodies and calibration standards were obtained from R&D Systems, Minneapolis, MN, USA (DuoSet® IC kits). MCF-7 cell protein lysates were isolated directly from the sensor chip at time points when changes in glycolysis and impedance occur. Two additional samples were also included: one containing cells treated for 24 h with 50 μM cisplatin and the other one treated with 10 μM for 24 h.

Preparation of cisplatin stock solution: We used 0.9% NaCl solution as a solvent for cisplatin. In general cisplatin is poorly soluble in this solvent. To obtain an homogenous cisplatin stock solution, we warm up the stock solution in the water bath at 37°C for 20 min during which we repeatedly vortex the solution. Otherwise we chose not to prepare a highly concentrated stock to prevent cisplatin precipitation. For the entire project we used 1 mM stock solution. Finally, because cisplatin is light sensitive, we covered the cisplatin stock solution with aluminum foil and switched off the clean bench lighting during pipetting and treatment.

3.5 RNA Isolation, DNA Microarray, Real-Time PCR

Total RNA was isolated directly from MCF-7 cells treated in the Bionas system at the time points when changes in glycolysis and impedance occurred using RNeasy Mini Kit (Qiagen, Hilden, Germany) according to the manufacturer's instructions. RNA quality was examined by agarose gel electrophoresis and the concentration was determined by UV absorbance. Affymetrix array analysis (GeneChip® U133 2.0plus Human Genome) was performed according to the manufacturer's protocol. Gene expression profiles were analyzed using the dChip software (Li and Wong 2001) and rank-based normalization (Kroll and Wölfl 2002); assignment of functional groups was done using DAVID Bioinformatics Resources 6.7 (Database for Annotation, Visualization, and Integrated Discovery) of NIAID (NIH) (Huang et al. 2009; Hosack et al. 2003) and free word data mining. For quantitative RT-PCR 250 ng of total RNA was reverse transcribed using oligo(dT)$_{18}$ primer (0.5 mg/mL, Fermentas) and random hexamers primer (100 μM, Fermentas) and RevertAid^{TM} Premium Enzyme Mix. Gene expression was assayed by quantitative real-time PCR on LightCycler® 480 (Roche Applied Science, Penzberg, Germany) using LightCycler® 480 SYBR Green I Master with 1 μL cDNA (1:5 dilution of transcribed cDNA) and the primers listed in the *Materials* section (2.5 Primers and Oligonucleotides).

3.6 Intracellular ROS Determination

MCF-7 and HT-29 cell lines (10^5 cells/mL medium) were treated with 50 μM cisplatin for 6, 10, 12, and 24 h. After treatment they were trypsinized and resuspended in 1 mL of Dulbecco's PBS (Gibco Invitrogen, Carlsbad, USA) supplemented with 1% bovine serum albumin (PAA Laboratories, Pasching, Austria). 10^5 cells were then incubated for 15 min in the dark with 10 μM dihydroethidium. Accumulation of reactive oxygen species (ROS) was measured by flow cytometry using a FACSCalibur instrument (Becton Dickinson, New Jersey, USA). Excitation and emission settings were 488 nm and 564–606 nm (FL2 filter), respectively. Results are presented as a percentage of cells that showed physiological levels of ROS (low ROS) and a percentage of cells with a high level of ROS.

4 Results

4.1 Respiration and Glycolysis – Striking Differences Between Cancer Cell Lines

The relative changes in respiration, glycolysis, and impedance of all tested cancer cell lines are compared in Fig. 4.1 – the basic glycolytic rate being lowest in MCF-7 and highest in HCT-116; respiration being highest in MCF-7 and lowest in HT-29. Differences in glycolytic and respiratory activity may reflect variations in a cell's ability to react to severe stress conditions. For instance, the lower basic glycolytic rate in MCF-7 cells may allow these cells to respond to severe stress conditions by inducing this metabolic pathway, while glycolysis can not be further induced in the other cell lines. Treatment with stress-inducing compounds like methyl methanesulfonate (MMS) and *tert*-butyl hydroperoxide (*t*BHP) also causes MCF-7 to increase glycolysis upon inhibition of respiration, while HT-29 increases respiration under stress conditions (Fig. 4.2).

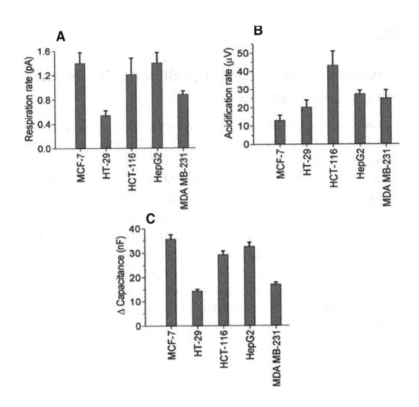

Fig. 4.1. Cell-type specific basal levels of metabolism and cell layer impedance.
Basal level of oxygen consumption, acidification of medium, and cell impedance
recorded in the cell biosensor chip without drug treatment and at the beginning of
experiments before treatment with cisplatin; **A**, basal respiratory activity (change of
oxygen of medium); **B**, basal glycolytic rate (change of extracellular pH); **C**, relative
difference of electrical capacity in comparison to cell-free biosensor chip (from
Alborzinia et al. 2011).

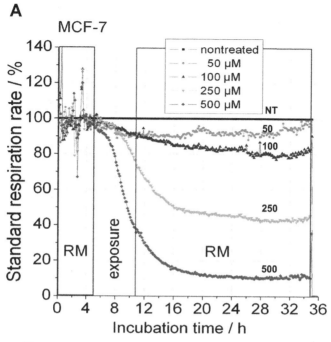

A

MCF-7

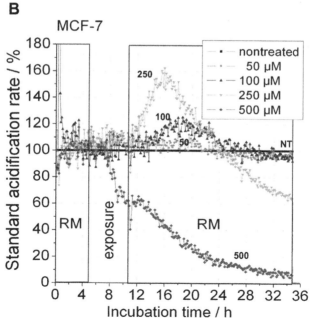

B

MCF-7

C

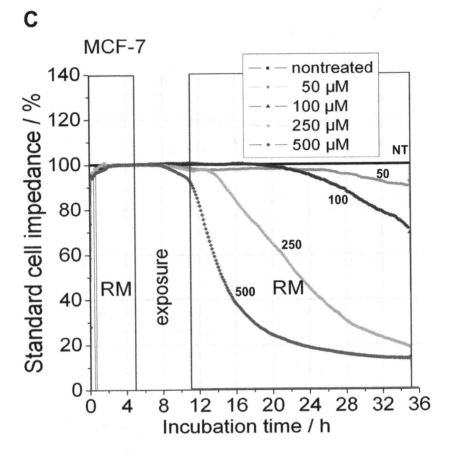

MCF-7

D

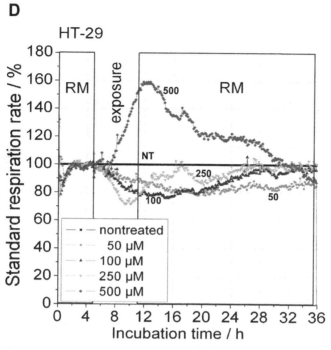

E

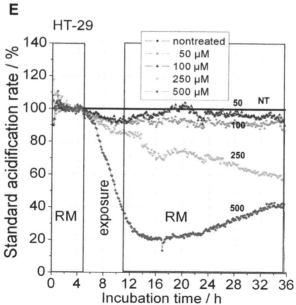

F

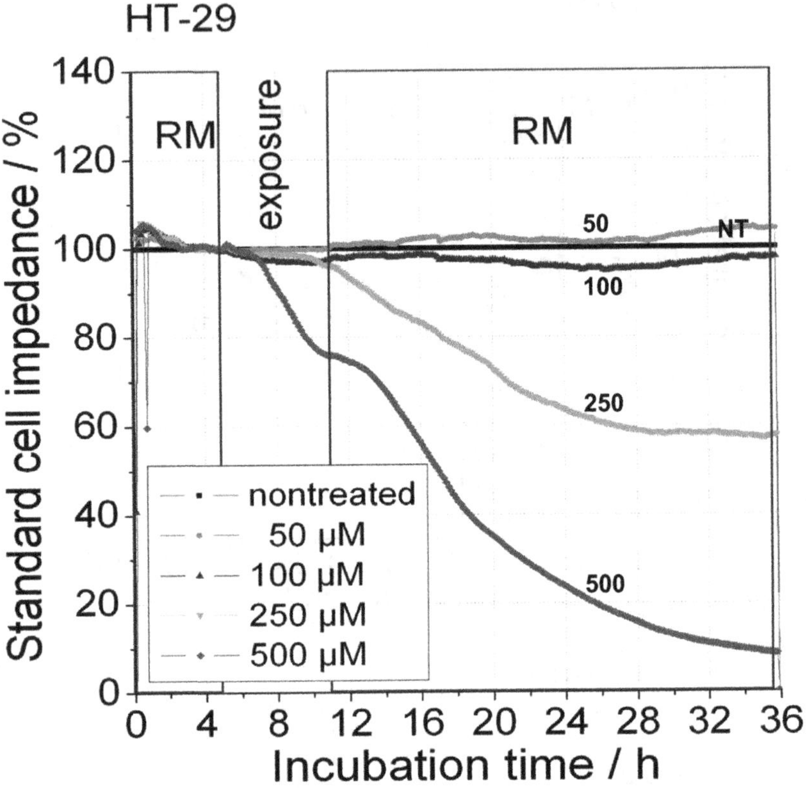
HT-29
RM
exposure
RM
50
NT
100
250
500
nontreated
50 µM
100 µM
250 µM
500 µM
Standard cell impedance / %
140
120
100
80
60
40
20
0
0
4
8
12
16
20
24
28
32
36
Incubation time / h

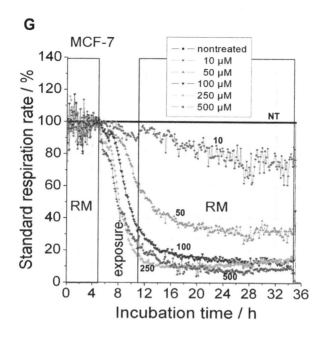

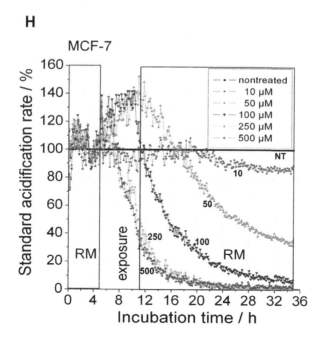

I

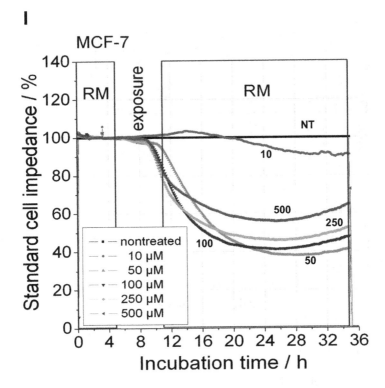

J

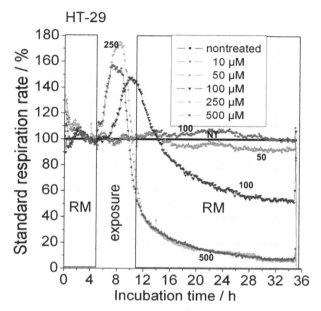

HT-29

Legend:
- nontreated
- 10 µM
- 50 µM
- 100 µM
- 250 µM
- 500 µM

Standard respiration rate / %

RM, exposure, RM

Incubation time / h

K

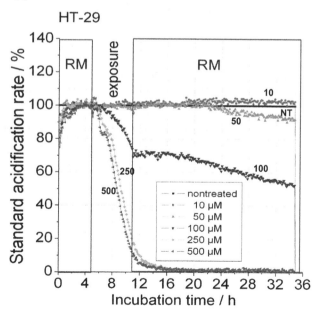

HT-29

RM, exposure, RM

Standard acidification rate / %

Legend:
- nontreated
- 10 µM
- 50 µM
- 100 µM
- 250 µM
- 500 µM

Incubation time / h

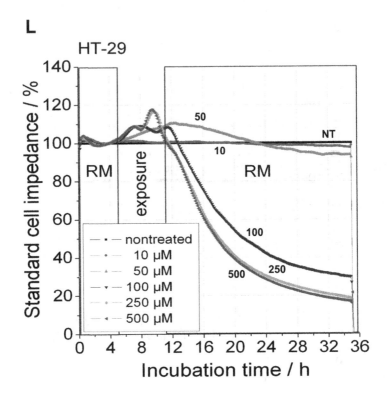

Fig. 4.2. Real-time response profiles of cellular respiration, glycolysis, and impedance to MMS and *t*BHP. Treatment started after 5 h of equilibration of cancer cell tissue cultures in the biosensor chip system and was continued over 6 h. *RM* (running medium) marks medium without added compound; *exposure,* marks time when substance was presented at indicated concentrations. **A, D, G, J,** change in respiration (oxygen in the medium), **B, E, H, K,** glycolysis, depicted as change in acidification of the medium (extracellular pH), **C, F, I, L,** change in impedance of the cell layer (potential between interdigitated electrodes) of MCF-7 (**A–C**: MMS and **G–I**: *t*BHP) and HT-29 (**D–F**: MMS and **J–L**: *t*BHP) (from Alborzinia et al. 2011).

4.2 Glucose Uptake Regulation – Online Measurements

To investigate the effect of glucose uptake regulation on cellular metabolism we first studied the impact of different glucose concentrations in the medium on cellular glycolysis and respiration rate. Base measurements were carried out using 1 g/L glucose, corresponding to a standard physiological blood glucose concentration. When glucose concentrations were lowered to 0.1 g/L first changes on respiration and glycolysis were recorded for HT-29 cells. Oxygen consumption increased under hypoglycemic conditions while glycolysis was reduced in a dose-dependent manner. No changes of impedance were recorded for any of the applied glucose concentrations, indicating that cellular morphology remained intact (Fig. 4.3).

A

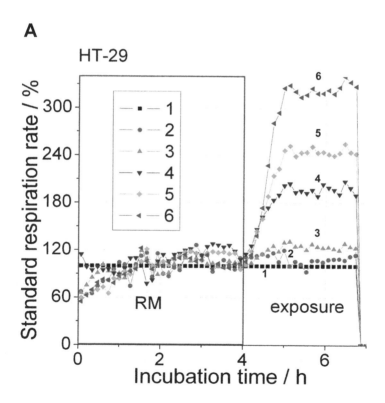

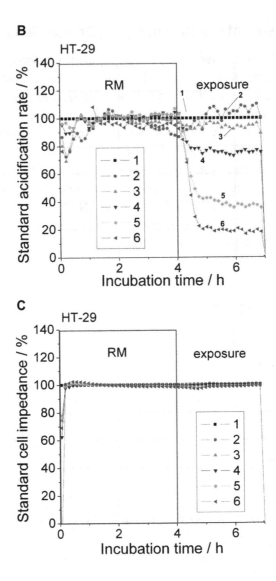

Fig. 4.3. Effect of glucose availability on respiration and glycolysis in HT29 cells. Concentration of glucose in the medium remained unchanged at 1 g/L glucose for the controls (**1,2**) or was reduced to 0.3 g/L (**3**), 0.1 g/L (**4**), 0.03 g/L (**5**), or 0.01 g/L glucose (**6**), respectively. The reduction of glucose led to **A**, a concentration-dependent increase of respiration and **B**, a proportional decrease of glycolysis, while **C**, cellular impedance remained unchanged.

The effect of glucose uptake on cellular metabolism was further investigated by means of two important glucose transporter inhibitors phloretin and phloridzin (*syn.* phlorizin). Application of 100 μM phlorizin to HT-29 cells led to an increase of respiration rate with a decrease of glycolysis; however, coincubation with phlorizin and phloretin caused respiration to become decreased. Accordingly, phloretin also reduced glycolysis, but only after a slight increase at the onset of treatment. Interestingly, however, as soon as treatment was terminated and also during the drug-free phase, respiration sharply increased and remained even higher than the level of nontreated cells and the glycolysis level reverted to the level of nontreated cells. So, both phloretin and phlorizin were shown to cause only transient inhibition of glucose transporters.

Impedance was slightly reduced by phlorizin, but strongly and immediately increased by phloretin – both effects being transient, as impedance went back to normal once treatment stopped (Fig. 4.4).

A

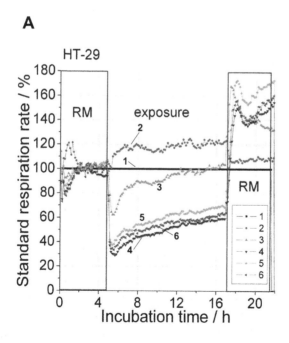

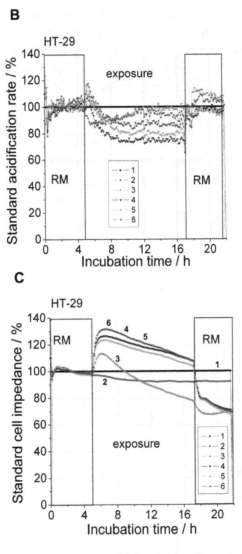

Fig. 4.4. Effects of glucose transport inhibitors phloretin and phlorizin on HT-29 cancer cells. Phlorizin alone (**2**) raised respiration and slightly reduced glycolysis, while phloretin overwrites the effect of phlorizin. Untreated control (**1**); 100 μM phlorizin (**2**), 100 μM phlorizin,and 10 μM phloretin (**3**), 100 μM phlorizin and 35 μM phloretin (**4**), 30 μM phlorizin and 35 μM phloretin (**5**), 35 μM phloretin (**6**); **A**: immediate reduction of respiration; **B**: immediate decrease when treated with phlorizin alone; when phloretin is present a short and transient increase of glycolysis, followed by a decrease, is observed; **C**, immediate increase in cellular impedance.

4.3 SIRT3 Overexpression – Respiration Profiles Measured Online

The effect of a single protein can be monitored for an extended period of time using this biosensor chip system. Sirtuin deacetylase SIRT3, a mitochondrial key regulatory enzyme of cellular metabolism, is over-expressed in HeLa cells and leads to an increase in respiration by up to 35% compared to cells transfected with the inactive SIRT3 mutant construct (Fig 4.5). The kinetics of SIRT3 overexpression was analyzed both by monitoring on the Bionas system and confirmed by western blot analysis (Fig. 4.5 B). Both, maximum SIRT3 overexpression (western blotting) and increase of cellular respiration (online analysis) were recorded at 18–20 h after SIRT3 transfection (Dransfeld et al. 2010); while SIRT3 overexpression had no effect on glycolysis and impedance (data not shown).

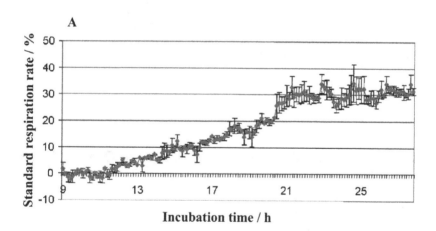

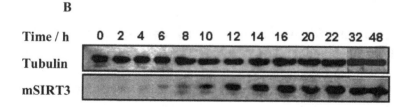

C

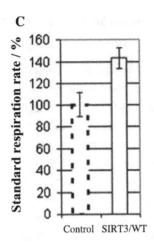

Fig. 4.5. SIRT3 overexpression – effect on cellular respiration. Online measure-ments **A**, SIRT3 overexpression increased the cellular respiration approximately 12 h post transfection compared with SIRT3 mutant. The maximum respiration level achieved around 21 h by 30–35% higher respiration level; **B**, Western blot analysis of HeLa cells overexpressing mitochondrial SIRT3 (mSIRT3-Flag). The first signal is visible 6 h post transfection and the maximum signal intensity achieved 16–20 h after transfection. An increase of SIRT3 expression correlate with an increase in respiration, presented from two independent experiments; **C**, respiration level 20 h after transfection with SIRT3 wt or controls transfected with the SIRT3 mutant (Dransfeld et al. 2010).

4.4 Cisplatin – Metabolic Changes Before Onset of Cell Death

Metabolic changes occurring under cisplatin treatment were monitored online using the same setup as described above for the cell lines MCF-7, MDA-MB-231, HT-29, HCT-116, and HepG2. After precultivation of the cells on the biosensor chip under standard tissue culture conditions in the CO_2 incubator, the biosensor chips were covered with a cell monolayer mounted in the flow cell. Experiments are started with equilibration to the biosensor chip conditions. Once signals for respiration, glycolysis, and impedance are stable over a several hours (4–5 h) treatment is started and continued over the given time period.

MCF-7 cells did not respond immediately to cisplatin treatment – only after about 5–6 h a first response to the treatment was noticed: a decrease in cellular respiration, an increase in glycolysis after 8–9 h at the highest concentration (50 μM), followed by a significant decrease in cellular impedance at 10–11 h after onset of cisplatin treatment (Fig. 4.6 A–C), the latter being considered as the onset of cisplatin-induced cell death. Higher cisplatin concentrations showed a more rapid and severe response. Interestingly, this response profile is very characteristic for MCF-7 cells – the colon carcinoma cells, HT-29 and HCT-116, reacted differently! While cisplatin treatment results in an immediate concentration-dependent decrease in cellular respiration in HCT-116 and a delayed decrease in the glycolytic rate (Fig. 4.6 G–I), HT-29 cells respond to cisplatin treatment (at least at lower concentrations) with a transient increase in respiration before cells start to die (Fig. 4.6 D–F). The transient increase of impedance in HCT-116 cells is also reproducible and may indicate a change in cell interactions in response to the treatment before induction of cell death (Fig. 4.6 G–I).

Treatment of HepG2 cells also shows an immediate decrease in respiration and a delayed decrease in glycolysis (Fig. 4.6 J–L). Again cells start to die independent of the applied cisplatin concentration after 8–9 h, but the extent of cell death is concentration dependent. An overview of the response time of the above cell lines to 50 μM cisplatin is presented in Table 4.1.

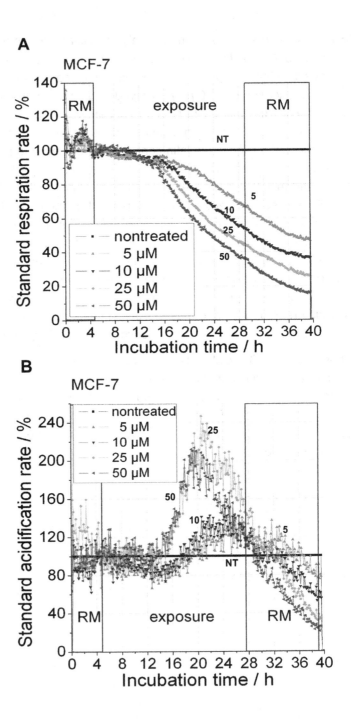

C

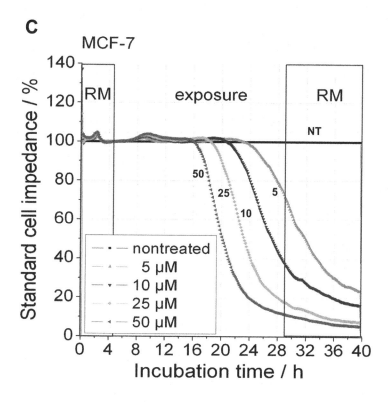

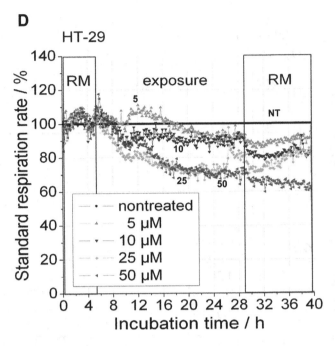

D

HT-29

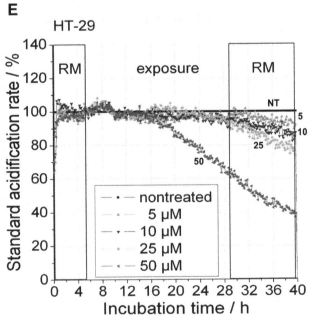

E

HT-29

F

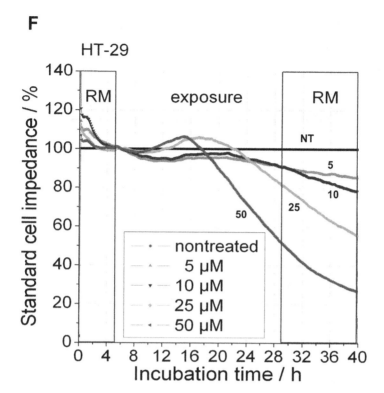

HT-29

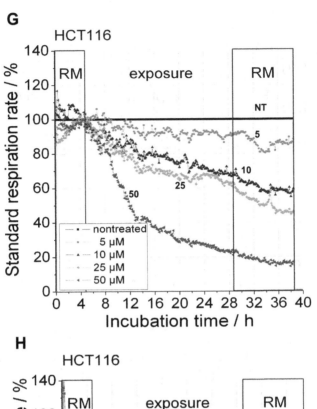

G

HCT116

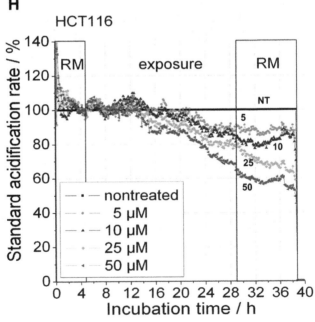

H

HCT116

I

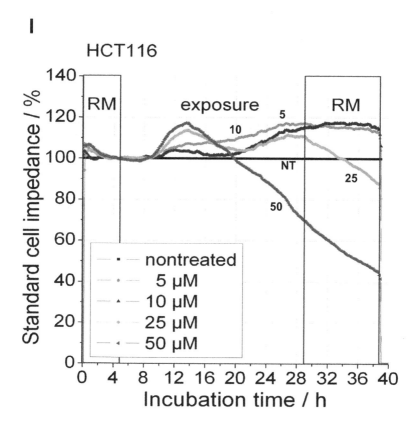

HCT116

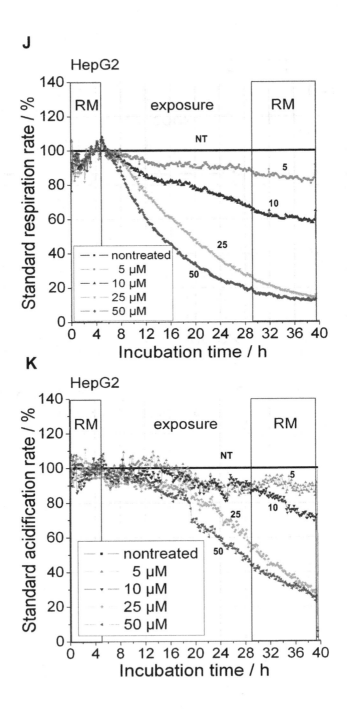

L

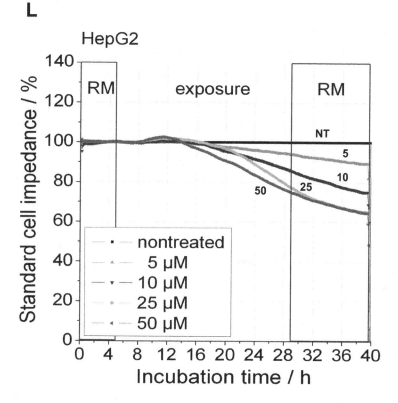

Fig. 4.6. Outline of the cell biosensor chip system and real-time response profiles of cellular respiration, glycolysis, and impedance upon cisplatin treatment. Cisplatin treatment started after 5 h of equilibration of cancer cell tissue cultures in the biosensor chip system and was continued over 24 h. RM (running medium) marks cisplatin-free medium; exposure marks time when cisplatin was presented at indicated concentrations (cisplatin-free control, 5, 10, 25, 50 μM). A, D, G, J, change in respiration (oxygen in the medium), B, E, H, K glycolysis, depicted as change in acidification of the medium (extracellular pH), C, F, I, L, change in impedance of the cell layer (potential between interdigitated electrodes) using four cancer cells lines, A, B, C, MCF-7, D, E, F, HT-29, G, H, I, HCT-116, and J, K, L, HepG2. Arrows indicate the time points of significant changes at the highest concentration (50 μM) of cisplatin used (from Alborzinia et al. 2011).

Table 4.1. Time points of metabolic and impedance changes in response to cisplatin (50 μM)*

Cancer cell line	Respiration change / hour	Glycolysis change / hour	Impedance change / hour
MCF-7	5–6 (Decrease)	8–9 (**Increase**)** 10–11 (Decrease)	10–11 (Decrease)
HT-29	Immediate response (Decrease)	9–10 (Decrease)	10–11 (Decrease)
HCT-116	Immediate response (Decrease)	6–7 (Decrease)	4–5 (**Increase**)* 8–9 (Decrease)
HepG2	Immediate response (Decrease)	6–7 (Decrease)	8–9 (Decrease)

* from Alborzinia et al. 2011
** denotes two transitions for one parameter: here, an increase precedes the final decrease of the measured value

Regarding the cisplatin-resistant breast cancer cell line MDA-MB-231 only a decrease in respiration can be observed at the highest (50 μM) cisplatin concentration, while cisplatin had no effect on glycolysis and cellular impedance, clearly indicating that these cells are not sensitive to the applied concentrations (Fig. 4.7 A–C). However, when the cisplatin concentration is significantly increased, also a clear cisplatin-induced cell death is noticed.

It should be emphasized, however, that none of the sensitive cell lines showed a recovery (even at lowest cisplatin concentrations) when cisplatin was removed from the medium after 24 h of treatment (Fig. 4.6) – indicating irreversible cellular damage. In addition, we also analyzed shorter exposure to cisplatin. Even after an exposure of 6 h, MCF-7 cells start to die in the same time frame and do not escape cisplatin-induced cell death (Fig. 4.8).

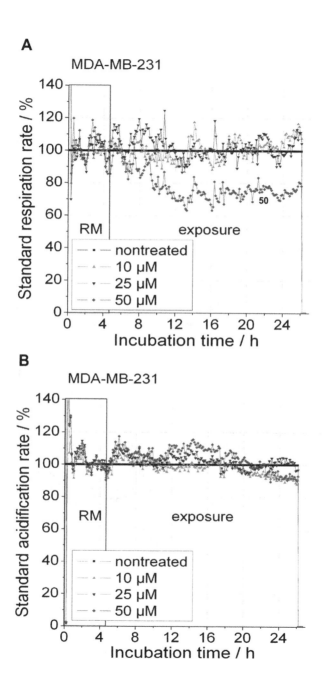

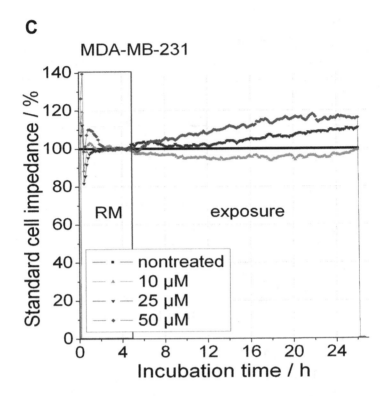

Fig. 4.7. Real-time response profiles of cellular respiration, glycolysis, and impedance upon cisplatin treatment of cisplatin-resistant breast cancer cell line MDA-MB-231. Cisplatin treatment started after 5 h of equilibration of cancer cell tissue cultures in the biosensor chip system and was continued over 20 h. RM (running medium) marks cisplatin free medium; *exposure* marks the time when cisplatin was presented at the indicated concentrations (cisplatin-free control, 10, 25, 50 μM). **A**, change in respiration (oxygen in the medium). **B**, glycolysis, depicted as change in acidification of the medium (extracellular pH). **C**, change in impedance of the cell layer (potential between interdigitated electrodes) using for cisplatin-resistant breast cancer cell lines MDA MB-231 (from Alborzinia et al. 2011).

A

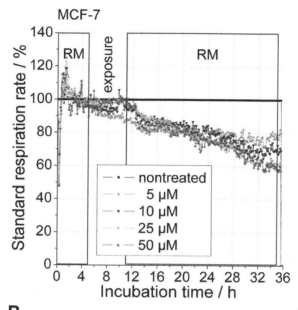

MCF-7

Standard respiration rate / %

RM — exposure — RM

— ∙ — nontreated
— ▲ — 5 µM
— ▼ — 10 µM
— ◆ — 25 µM
— ◄ — 50 µM

Incubation time / h

B

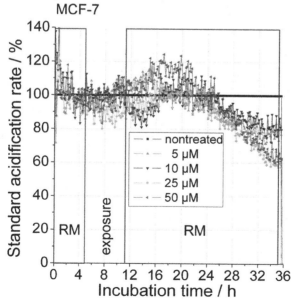

MCF-7

Standard acidification rate / %

— ∙ — nontreated
— ▲ — 5 µM
— ▼ — 10 µM
— ◆ — 25 µM
— ◄ — 50 µM

RM — exposure — RM

Incubation time / h

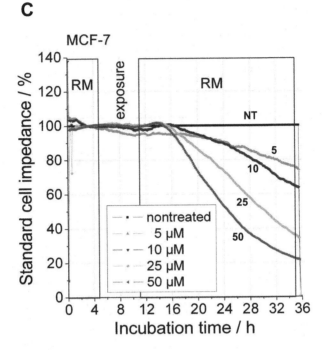

Fig. 4.8. Outline of the cell biosensor chip system and real-time response profiles of cellular respiration, glycolysis, and impedance of MCF-7 upon 6-h cisplatin treatment. Cisplatin treatment started after 5 h of equilibration of cancer cell tissue cultures in the biosensor chip system and was continued for 6 h. RM (running medium) marks cisplatin-free medium; exposure marks time when cisplatin was presented at indicated concentrations (cisplatin-free control, 5, 10, 25, 50 μM). **A,** change in respiration (oxygen in the medium), **B,** glycolysis, depicted as change in acidification of the medium (extracellular pH), **C,** change in impedance of the cell layer (potential between interdigitated electrodes).

4.4.1 Mitochondrial Oxygen Consumption

Because of the immediate effect of cisplatin on respiration of colon cancer and liver cancer cell lines, we decided to analyze whether cisplatin directly affects mitochondrial respiration using a mitochondrial activity assay with isolated mouse liver mitochondria (Fig. 4.9). While the well-established mitochondrial inhibitor rotenone and the uncoupler CCCP clearly blocked and increased mitochondrial activity, respectively, cisplatin had no effect on mitochondrial respiration within 4–5 h, independent of the cisplatin concentration used. This clearly shows that respiratory changes observed with intact cells in the cell biosensor system must have resulted from a cellular response to cisplatin treatment.

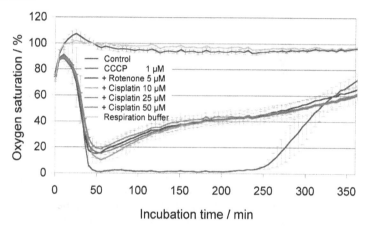

Fig. 4.9. Respiration of isolated mouse liver mitochondria. Respiration consumes oxygen, depicted as a decrease of oxygen saturation over time. After about 50 min mitochondrial respiration buffer is exhausted, leading to continuous reduction of respiratory activity. Inhibition of mitochondrial activity blocks oxygen consumption, resulting in continuous high oxygen concentration: (*i*) control (*purple*): mitochondria with respiration buffer, but without cisplatin; (*ii*) CCCP (*brown*): carbonyl cyanide 3-chlorophenylhydrazone uncouples respiration leading to maximum consumption and complete depletion of oxygen (0%); (*iii*) rotenone (*dark blue*): an inhibitor of respiratory chain complex I, completely blocks mitochondrial respiration leaving oxygen concentration unchanged at 100%; (*iv*) respiration buffer (*yellow*): no mitochondria, no cisplatin. In all three cisplatin samples, the drug was added at the start of measurement at the indicated concentrations (10, 25, 50 μM): there is no apparent difference between the control (*purple*) and cisplatin treatment (from Alborzinia et al. 2011).

We therefore looked at the formation of ROS in response to cisplatin treatment after 6, 10, 12, 24 h in MCF-7 and HT-29 cells, assuming that ROS could be the critical mediator of early cisplatin response, including respiration change, and also an important mediator of cisplatin-induced cell death (Simon et al. 2000). Surprisingly, we did not detect significant ROS formation in MCF-7 and HT-29 cells treated with 50 μM cisplatin at indicated time points (Fig. 4.10).

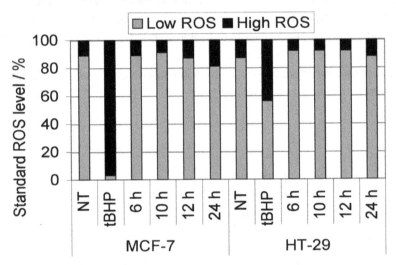

Fig. 4.10. ROS levels upon cisplatin treatment. MCF-7 and HT-29 were treated with 50 μM cisplatin for the indicated incubation periods (6, 10, 12, 24 h). As positive control, cells were treated with 500 μM tBHP (tert-butyl hydroperoxide) for 3 h. No significant induction of intracellular ROS with 50 μM cisplatin was observed at the analyzed time points (from Alborzinia et al. 2011).

4.4.2 Cisplatin – Cell Signaling Profiles

At those time points at which the most significant changes were identified by online measurement we performed a more detailed analysis of signal transduction in the cisplatin-sensitive breast cancer cell line MCF-7. In particular, we recorded the level of phosphorylation of selected signal transduction mediators connected with cellular proliferation. Protein lysates were performed from MCF-7 cells directly obtained off the sensor chip at those particular time points when changes in glycolysis and impedance occurred.

To further investigate the mechanisms underlying the observed changes in cellular metabolism and onset of cell death upon cisplatin treatment we analyzed changes in cell signaling in MCF-7 cells at the time points when changes in glycolysis and impedance occur. The phosphorylation of key signal transduction mediators was measured by measns of a phosphoprotein ELISA microrray (Fig. 4.11).

Hardly any change was noticeable in the phosphorylation of most analyzed signaling proteins [(phosphoproteins: GSK-3ß(S9), Akt1(S473), ERK1(T202/Y204), ERK2(T185/Y187), and CREB (S133)] at the indicated time points when glycolysis (~8 h) and impedance (~11 h) changed. In treated samples a significant decrease in p-GSK-3ß and p-Akt1 occurred after 24 h. In particular using samples from untreated controls at start and end of experiment, it is clear that no change occurs in untreated cells. This decrease was clearly visible in two different treatment regimes. Using the same strong cytotoxic concentration of cisplatin (50 μM) for 10–11 h (the time point of impedance change) followed by 14 h in drug-free medium, not onlyp-GSK-3ß and p-Akt1 were significantly decreased, but also an increase in the concentration of p-ERK1 was observed. Interestingly, continuous treatment with a lower dose of cisplatin (10 μM) for 24 h also showed a significant but lower reduction in p-GSK-3ß and p-Akt1, without any change in p-ERK1. Both p-Akt1 and p-GSK-3ß are involved in cell cycle regulation (Panka et al. 2008) and reduced p-Akt1 indicates cell cycle arrest and reduced anti-apoptotic activity (Datta et al. 1999).

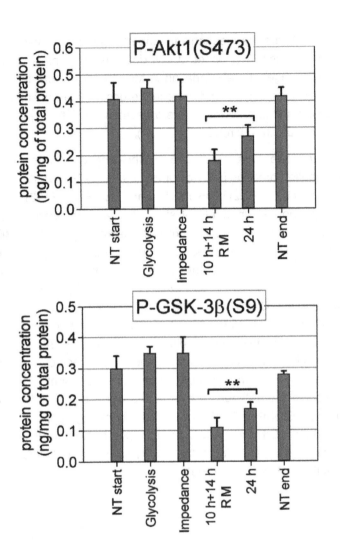

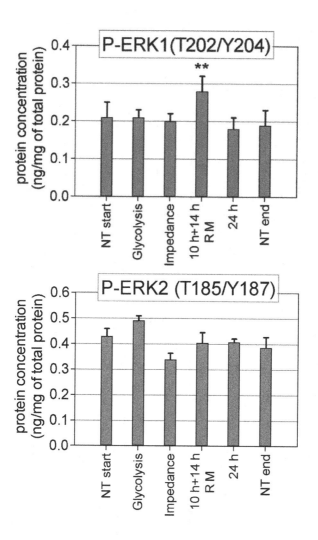

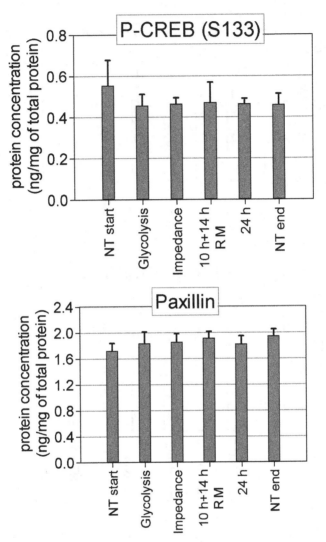

Fig. 4.11. Change in the pathway activities of Akt1 (A), GSK-3ß (B), ERK1 (C), ERK2 (D), P-CREB (E), paxillin (F). Concentrations of selected phosphoproteins were measured with an ELISA microarray in MCF-7 cells at the time points when glycolysis and impedance were significantly changed with 50 μM cisplatin treatment, and after 24 h with a 10-h pulse with 50 μM cisplatin (followed by 14 h with cisplatin-free medium [RM]) or after continued treatment for 24 h with 10 μM cisplatin. For control, the concentration of phosphoproteins was measured in nontreated (*NT start*) cells at the same time points when glycolysis changed in treated samples, and also at the end of experiments (24 h, *NT end*) (from Alborzinia et al. 2011).

4.4.3 Cisplatin – Gene Expression Profiles

To obtain a more detailed picture of the cellular response of MCF-7 cells to cisplatin, we also analyzed gene expression profiles using whole genome microarrays (Affymetrix GeneChips®). For this, treatment was performed in the cell biosensor chip system and cells were lysed and RNA was collected directly on the biosensor chip immediately after the first change in (*i*) glycolytic rate and (*ii*) impedance became visible when treated with 50 μM cisplatin. For control, cells were incubated in parallel in another biosensor chip unit without addition of cisplatin and collected for RNA preparation at a corresponding time point. For data analysis the cutoff was a minimal 1.75-fold change in one of the two time points and a minimal normalized signal of 100 for the higher value. With this cutoff we obtained 1338 significantly regulated probe sets, of which 335 were up- and 1003 down-regulated (Supplementary Table, see Appendix).

Most genes were differentially expressed at both time points, but a significant number was stronger regulated at the later time point of impedance change. Top-regulated genes for each treatment are shown in Table 4.2. While top-regulated (strongly regulated) genes at glycolysis change were not further induced, the according list (Table 4.2) from impedance change shows that gene expression changed significantly more within the additional (about) 2 h before cells began to die (impedance change). The reliability of the gene expression profiles was confirmed by quantitative real-time RT-PCR analysis of selected genes in independent experiments (Fig. 4.12).

Table 4.2. Genes regulated by cisplatin in MCF-7 at change of (**A**, **B**) glycolysis (8–9 h) and (**C**, **D**) impedance (10–11 h)

A) Up-regulated genes at time point of glycolysis change

		fold change glycolysis	fold change impedance	imp/gly
1	HSPA6	9.61	12.35	1.29
2	EIF1	4.77	8.90	1.87
3	HEXIM1	4.00	5.90	1.47
4	PMAIP1	4.00	5.79	1.45
5	FLJ44342	4.24	5.22	1.23
6	Hs,155364,0	3.81	4.51	1.18
7	ATF3	3.14	4.14	1.32
8	ETS2	3.05	4.09	1.34
9	PHLDA1	2.87	3.83	1.33
10	RGS2	2.93	3.74	1.28
11	LOC284801	4.22	3.58	0.85
12	Hs,41272,0	2.74	3.29	1.20
13	C1orf63	2.69	3.19	1.19
14	GADD45A	2.78	3.04	1.10
15	SERTAD1	2.92	3.04	1.04

B) Down-regulated genes at time point of glycolysis change

		fold change glycolysis	fold change impedance	imp/gly
1	FAM125B	-13.85	-16.85	1.22
2	ZNRF3	-9.35	-7.14	0.76
3	GPR39	-7.44	-12.38	1.66
4	MAML2	-5.71	-7.26	1.27
5	ARHGAP24	-5.61	-7.34	1.31
6	RIN2	-5.16	-9.02	1.75
7	TCF7L2	-5.08	-7.08	1.39
8	FGD4	-5.07	-5.69	1.12
9	TNS3	-4.89	-5.49	1.12
10	Hs,60257,0	-4.76	-4.90	1.03
11	GLI3	-4.68	-5.04	1.08
12	FCHSD2	-4.64	-3.75	0.81
13	HS6ST3	-4.59	-5.50	1.20
14	Hs,323099,0	-4.58	-3.61	0.79
15	Hs,235857,0	-4.48	-4.42	0.99

C) Up-regulated genes at time point of impedance change

		fold change glycolysis	fold change impedance	imp/gly
1	HSPA6	9.61	12.35	1.29
2	EIF1	4.77	8.90	1.87
3	HBA, a1/a2	1.92	6.36	3.31
4	MAPK12	2.04	6.07	2.97
5	ZNF277	2.46	6.07	2.47
6	HEXIM1	4.00	5.90	1.47
7	PMAIP1	4.00	5.79	1.45
8	FLJ44342	4.24	5.22	1.23
9	ID2, ID2B	2.34	5.00	2.14
10	Hs,155364,0	3.81	4.51	1.18
11	Hs,102981,0	2.65	4.30	1.62
12	ATF3	3.14	4.14	1.32
13	C17orf91	2.49	4.12	1.66
14	ETS2	3.05	4.09	1.34
15	Hs,14613,0	2.43	4.00	1.65

D) Down-regulated genes at time point of impedance change

		fold change glycolysis	fold change impedance	imp/gly
1	FAM125B	-13.85	-16.85	1.22
2	GPR39	-7.44	-12.38	1.66
3	SRGAP1	-3.68	-10.74	2.92
4	SAP30	-2.38	-10.66	4.48
5	RAB28	-2.99	-9.45	3.16
6	RIN2	-5.16	-9.02	1.75
7	FGFR2	-4.29	-7.84	1.83
8	TMTC2	-3.46	-7.80	2.25
9	ARHGAP24	-5.61	-7.34	1.31
10	MAML2	-5.71	-7.26	1.27
11	ZNRF3	-9.35	-7.14	0.76
12	TCF7L2	-5.08	-7.08	1.39
13	SYT1	-3.98	-7.02	1.76
14	PCDH7	-4.04	-6.80	1.68
15	LYN	-3.65	-6.74	1.84

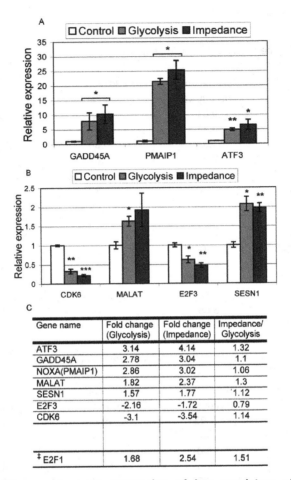

A

Relative expression

☐ Control ▨ Glycolysis ■ Impedance

GADD45A   PMAIP1   ATF3

B

Relative expression

☐ Control ▨ Glycolysis ■ Impedance

CDK6   MALAT   E2F3   SESN1

C

Gene name	Fold change (Glycolysis)	Fold change (Impedance)	Impedance/ Glycolysis
ATF3	3.14	4.14	1.32
GADD45A	2.78	3.04	1.1
NOXA(PMAIP1)	2.86	3.02	1.06
MALAT	1.82	2.37	1.3
SESN1	1.57	1.77	1.12
E2F3	-2.16	-1.72	0.79
CDK6	-3.1	-3.54	1.14
‡ E2F1	1.68	2.54	1.51

Fig. 4.12. Changes in gene expression of key regulators at the time of glycolysis and impedance change in MCF-7. To confirm large-scale gene expression profiling data, selected genes with a focus on p53 signaling were analyzed by quantitative real-time RT-PCR using independent samples. **A,** real-time RT-PCR shows strong induction of GADD45A, PMAIP1 and ATF3; and **B,** low, but significant regulation of MALAT and SESN1 (up) and CDK6 and E2F3 (down). *Controls*: samples from incubation on biosensor chip without cisplatin treatment; *Glycolysis*: samples from biosensor chip at time of glycolysis change; *Impedance*: samples from biosensor chip at time of impedance change. **C,** table summarizing results from gene expression profiling for the same genes. ‡E2F1 is included because it shows similar significance in micro array data as E2F3. Fold change relative to control sample for glycolysis and impedance change. Impedance/Glycolysis: relative ratio of changes in glycolysis and impedance. Significance values: * $p < 0.05$; ** $p < 0.01$; *** $p < 0.001$ (from Alborzinia et al. 2011)

Further analysis by gene ontology and pathway annotation using DAVID at NIAID (NIH) (Huang et al. 2009; Hosack et al. 2003) showed an interesting enrichment of genes suggesting a specific pro-apoptotic transition. Genes up-regulated at change of impedance were (among other less well-defined groups) clearly enriched for p53-signaling ($p = 0.0023$), cell cycle control (cyclin-dependent protein kinase activity: $p = 0.001$), and programmed cell death (apoptosis: $p = 0.0058$) (Table 4.3).

Table 4.3. Significantly regulated functional groups

Induced (impedance/all)

functional group (pathway)	genes	p-value
response to protein stimulus	5	0.00063
regulation of cyclin-dependent protein kinase activity	4	0.001
heparin-binding	4	0.0012
p53 signaling pathway	4	0.0023
apoptosis	8	0.0058

Down (impedance/all)

pathway	genes	p-value
Wnt signaling pathway	16	0.000008
ErbB signaling pathway	11	0.000082
MAPK signaling pathway	17	0.0016
TGF-beta signaling pathway	9	0.0019
Hedgehog signaling pathway	7	0.0033
VEGF signaling pathway	8	0.0034
Adherens junction	8	0.0039
Aldosterone-regulated sodium reabsorption	6	0.0041
Insulin signaling pathway	10	0.0089

Induced (first response/glycolysis)

functional group (pathway)	genes	p-value
response to protein stimulus	4	0.00058
regulation of cyclin-dependent protein kinase activity	3	0.0031
stress response	3	0.0039
response to organic substance	6	0.0042

assignments/significance values obtained with DAVID Functional Annotation Tool, National Institute of Allergy and Infectious Diseases (NIAID) NIH

Down-regulated genes were enriched for various cell proliferation-associated cell processes, in particular, wnt ($p = 0.000008$), tgfß ($p = 0.0019$), MAP-kinase ($p = 0.0016$), hedgehog ($p = 0.0033$), and erbb ($p = 0.000082$) signaling (Table 4.2). The pro-apoptotic regulation is well reflected in the main regulatory circuit of p53 control, with up-regulation of E2F1, but down-regulation of E2F3 (Fig. 4.12 C). Further analysis of genes already significantly induced at glycolysis, suggests activation of a general stress response including cell cycle arrest, before induction of cell death (response to protein stimulus, regulation of cyclin-dependent protein kinase activity, stress response, response to organic substance; $p < 0.05$). The complex regulatory activities, initial stress response, and transition to apoptosis are also reflected in the strong induction of p53 signaling mediators (Fig. 4.12), namely ATF3 a mediator of p53 signaling, the pro-apoptotic protein NOXA(PMAIP1), and the cell cycle regulator GADD45A, as well as the down-regulation of CDK6, required for cell cycle progression (Fig 4.13).

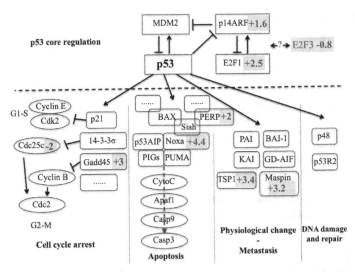

Fig. 4.13. Pathways of the p53 control circuit. p53 controls downstream regulatory signaling towards cell cycle arrest, apoptosis, metastasis, and DNA damage or repair. The core regulatory unit consist of p53, MDM2, p14ARF and E2f1; p53 and MDM2 are effected by p14ARF, E2F1, and E2F3; downstream signaling pathways involving Gadd45, Cdc25c, Noxa, PERP, TSP1, Maspin are then controlled by p53 (fold changes detected in gene expression microarray experiments are indicated for the respective genes; modified after Harris and Levine 2005)

Although respiration in intact cells was immediately inhibited by cisplatin, no instant effect on mitochondrial activity was observed within that particular time frame. There was a significant difference between cellular responses among tested cell lines, resulting from the pronounced difference between basic glycolytic and respiration levels. Also, changes in prosurvival and proliferation signaling are not apparent until the onset of cell death. Only after cells started to die, a decrease of survival signaling Akt/GSK-3ß and proapoptotic signaling (ERK1) appear. Nevertheless, the primary p53 regulatory circuit is clearly modulated towards MDM2/p53 DNA damage response at the point of impedance reduction.

4.5 Organometallics – Real-Time Analysis of Metabolic Responses in Cancer Cells

The powerfulness of the presented cisplatin activity measurements shows that cell biosensor chips can also be used for additional screening of the biological activity of new drug candidates.

4.5.1 Ruthenium(II) Polypyridyls

For further understanding of cellular metabolic response upon exposure to cytotoxic compounds we have used several organometallic compounds and analyze cellular metabolic response in real time. In this regard we have studied the mode of action of ruthenium(II) polypyridyl compounds (complexes **1–5**) on cellular metabolism and morphology of HT-29 colon cancer cells. We exposed HT-29 cells to each of the five compounds at 100 μM for 24 h using the Bionas 2500 system (Schatz-schneider et al. 2008) (Fig. 4.14). The glycolysis rate of HT-29 cells treated with ruthenium compounds **1–5**, as shown in Fig. 4.15, slightly decreases during the exposure in a similar way. However, cells treated with compounds **1–4** somewhat recover during the drug-free phase and after 24 h of treatment. Dppn complex **5** has an irreversible effect on cellular glycolysis activity, therefore the acidification rate continues to decrease during the drug-free period.

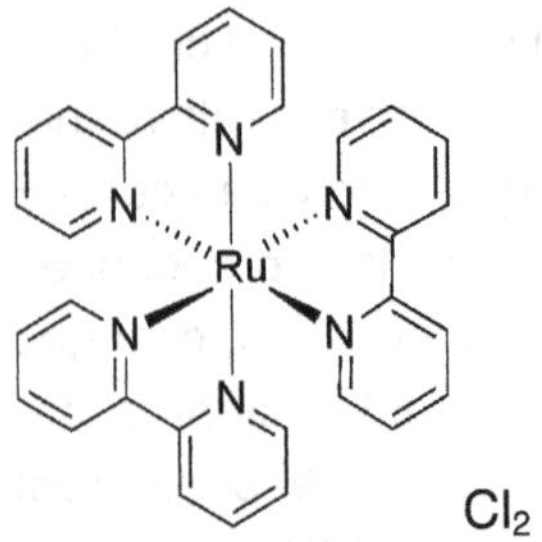

Fig. 4.14. Ruthenium(II) polypyridyl complexes

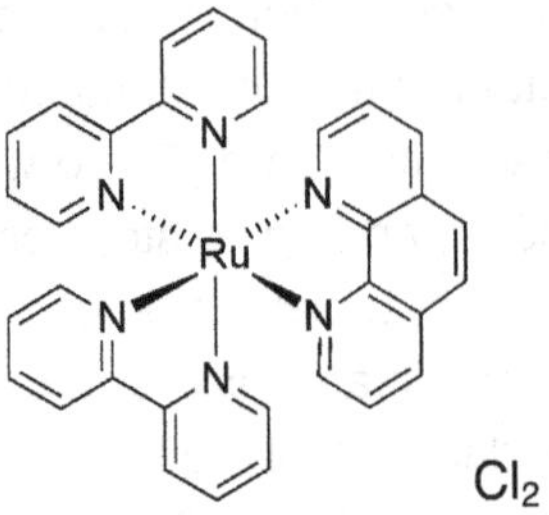

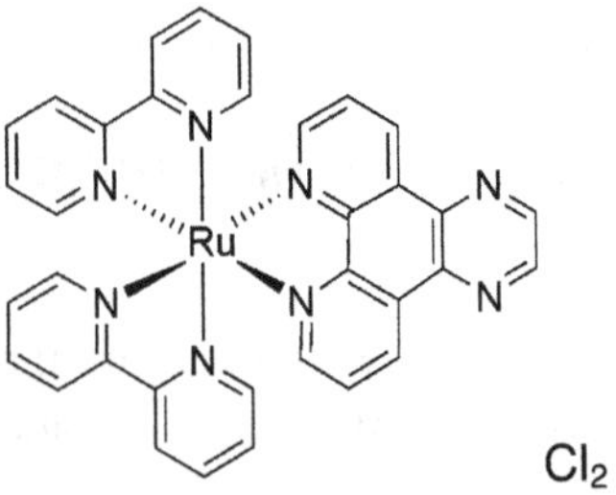

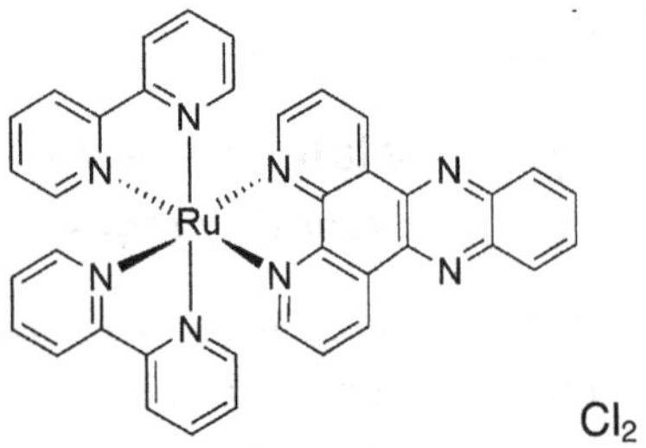

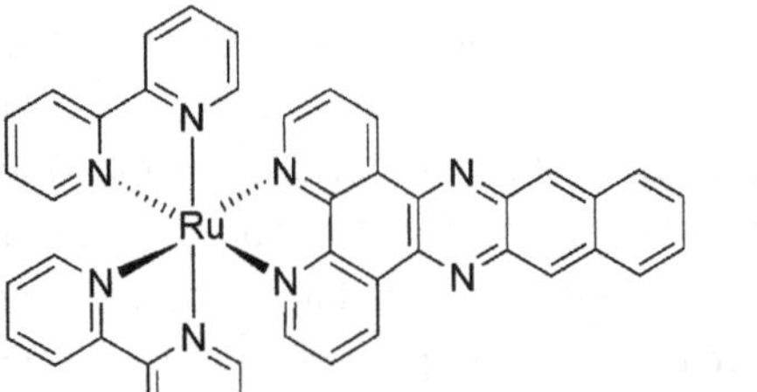

Oxygen consumption rate which is an indicator of mitochondrial activity and cellular respiratory rate, is only slightly affected by ruthenium(II) polypyridyl compounds (1–5). Finally, the impedance of the cell layer growing on the chip surface is influenced by the insulating properties of the cell membrane. As shown in Fig. 4.15, the HT-29 cells clearly respond to all five ruthenium(II) polypyridyl compounds (**1–5**) by changes in cell impedance. Whereas the effects of compounds **1–4** on HT-29 cells are very small, a dramatic decrease in cell impedance by almost 70% is observed for dppn complex **5**. This is clear indication of cellular morphological changes or cell–cell and cell–matrix contact. In the drug-free step the effect of complex **5** is irreversible for HT-29 cells, since the impedance hardly recovered when the cells were treated with medium without compound. The data, however, do not allow us to distinguish between cell–cell and cell–matrix changes nor whether the compound has a direct effect on the cell membrane.

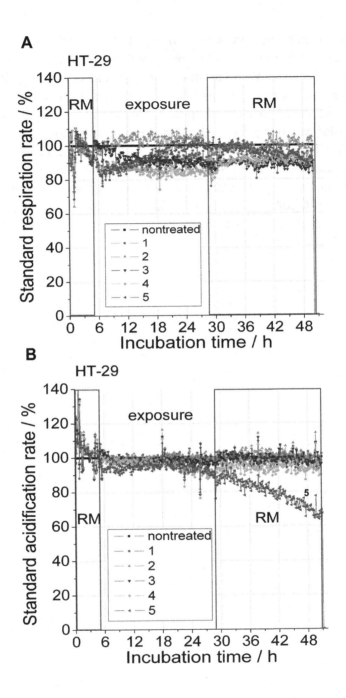

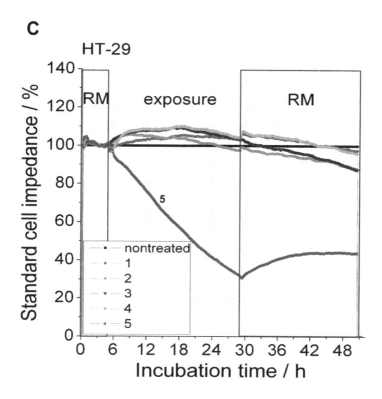

Fig. 4.15. Ruthenium(II) polypyridyl complexes. Real-time response profiles of cellular respiration, glycolysis, and impedance of HT-29. Treatment started after 5 h of equilibration phase and was continued for 24 h. *RM* (running medium) marks compounds free medium; *exposure* marks the time when compounds (1–5) were presented at the indicated concentration (100 µM). **A**, change in respiration (oxygen in the medium); **B**, glycolysis, depicted as change in acidification of the medium (extracellular pH); **C**, change in impedance of the cell layer (potential between interdigitated electrodes) (Schatzschneider et al. 2008).

4.5.2 Rhodium(III) Polypyridyls

The highly cytotoxic rhodium(III) polypyridyl complexes (**6** and **7**) were studied in order to determine the response of HT-29 and MCF-7 cells in terms of cellular metabolism and morphological changes by using the Bionas 2500 system (Harlos et al. 2008) (Fig. 4.16). The oxygen consumption of both HT-29 and MCF-7 cells is strongly affected during the first hours of exposure to 5 μM of the two compounds (Fig. 4.17 A–F). In contrast to MCF-7, HT-29 cells exposed to 2 μM of **6** and 2 or 5 μM of **7** showed a certain degree of recovery during the drug-free medium phase after a rapid initial decrease to a 60% level was observed. Regeneration is no longer visible for 5 μM of **6**, possibly due to permanent cellular damage. MCF-7 cells showed no recovery for both **6** and **7** at 2 or 5 μM concentrations. Interestingly, HT-29 cells showed a short transient increase at the beginning of the treatment, which was not seen for MCF-7 cells. After small positive and negative fluctuations in oxygen consumption the signal values approached the control levels during the drug-free step (Fig. 4.17 A).

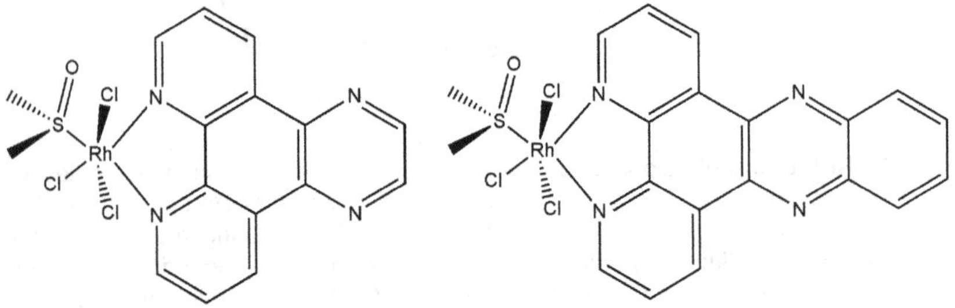

6   **7**

Fig. 4.16. Trichloridorhodium(III) polypyridyl complexes

Extracellular acidification, which is due to changes in the glycolytic pathway, was strongly affected by 2 or 5 μM of **6** and **7** in both HT-29 and MCF-7 cells. The glycolytic pathway in HT-29 cells is more sensitive to these compounds as compared to MCF-7 cells (Fig. 4.17 B, E).

A significant dose-dependent decrease in the extracellular acidification rate is apparent for HT-29 cells treated with **6** (Fig. 4.17 B) preceding the drug-free phase. Interestingly, glycolysis increased in MCF-7 cells, in contrast to HT-29 cells, upon treatment with both **6** and **7**. But nevertheless, glycolysis levels decreased during the drug-free phase in MCF-7, which showed a permanent cellular damage upon 6 h of exposure to both **6** and **7** (Fig. 4.17 E). The impact of **7** is much weaker for both cell lines, in particular for MCF-7, where the acidification rate is close to 100% after 28 h. As expected, impedance, which reflects cellular morphological changes and cellular viability, was affected strongly in both HT-29 and MCF-7 cells with both **6** and **7**, but **6** showed higher cytotoxity (Fig. 4.17 C, F). Moreover, MCF-7 cells appeared to be more sensitive to **6** and **7** in comparison to HT-29 cells. But neither HT-29 nor MCF-7 were able to recover during the drug-free phase, indicating a permanent cellular damage due to **6** and **7** treatment.

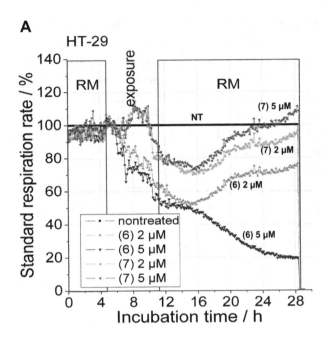

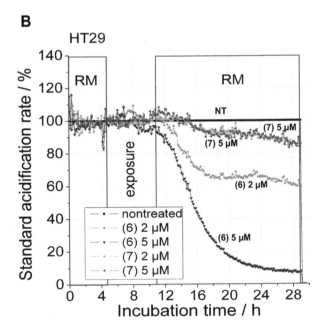

c

HT-29

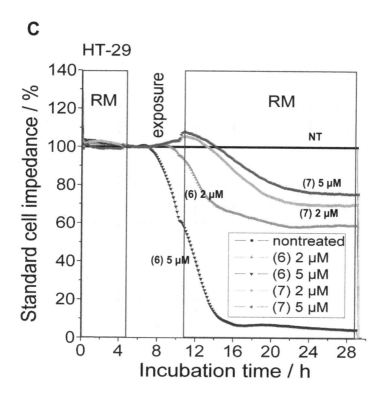

D

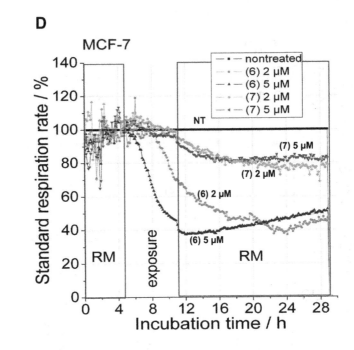

E

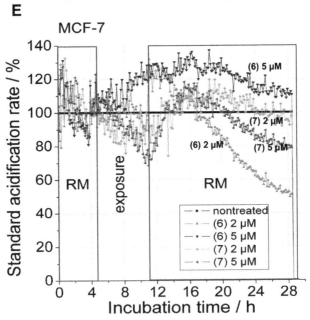

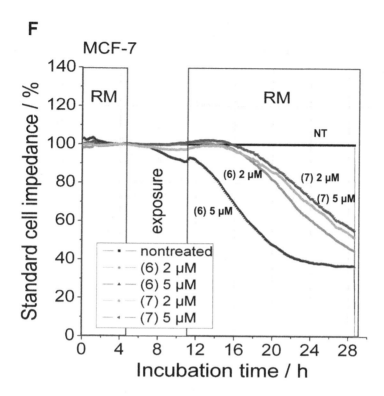

Fig. 4.17. Rhodium(III) polypyridyl complexes. Real-time response profiles of cellular respiration, glycolysis, and impedance of HT-29 and MCF-7. Treatment started after 5 h of equilibration and was continued for 6 h. *RM* (running medium) marks compounds free medium; *exposure* marks the time when compounds (**6–7**) were presented at the indicated concentrations (2 and 5 μM). **A** and **D,** change in respiration (oxygen in the medium); **B** and **E,** glycolysis, depicted as change in acidification of the medium (extracellular pH); **C** and **F,** change in impedance of the cell layer (potential between interdigitated electrodes) (Harlos et al. 2008).

4.5.3 Phenylenediamine Iron

Changes in cellular metabolism and morphology of MCF-7 cells in response to compound **8** (metal complex) was investigated using the Bionas 2500 system (Hille et al. 2009) (Fig. 4.18). In our investigations this compound had a significant and immediate effect on cell metabolism (Fig. 4.19). Glycolysis immediately decreased in all applied drug concentrations (1, 2.5, and 5 μM), while cellular respiration increased sharply at the onset of exposure as an immediate stress response, but decreased 3–4 h from the onset of exposure to **8**. Cellular impedance, decreased rapidly from about 1.5–3 h after onset of drug administration which indicates the induction of cells. Depression of respiration, glycolysis, and cellular impedance even continued during the drug-free phase, which is clearly indicative of irreversible cellular damage of MCF-7 cells by this metal complex.

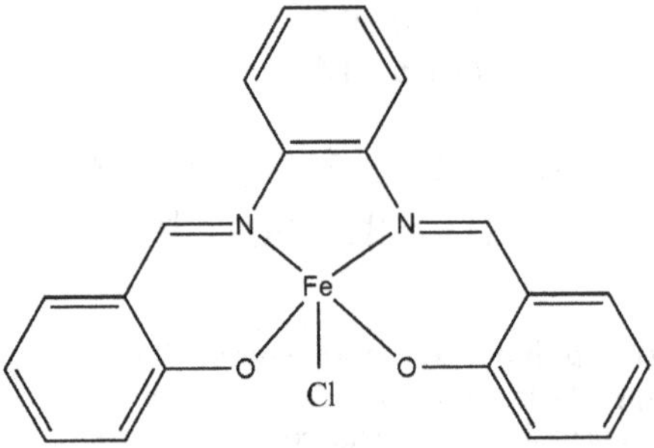

8

Fig. 4.18. [Salophene]iron complex

86

A

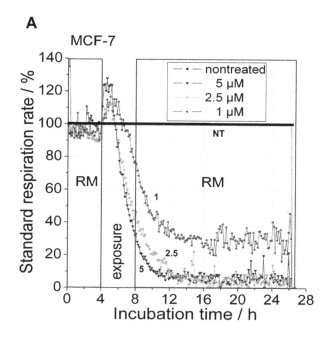

B

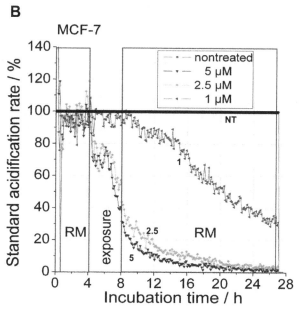

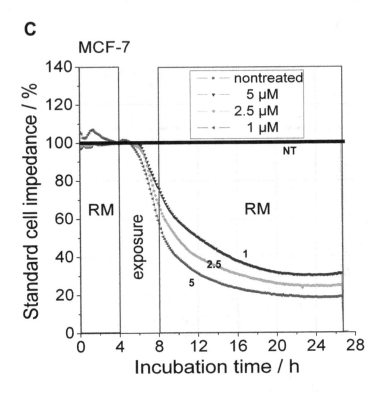

Fig. 4.19. [*N,N'*-Bis(salicylidene)-1,2-phenylenediamine]–iron complex.
Real-time response profiles of cellular respiration, glycolysis, and impedance of MCF-7. Treatment started after 4 h of equilibration for 4 h. *RM* (running medium) marks compounds free medium; *exposure* marks the time when compound **8** was presented at the indicated concentrations (1, 2.5, and 5 μM). **A**, change in respiration (oxygen in the medium); **B**, glycolysis, depicted as change in acidification of the medium (extracellular pH); **C**, change in impedance of the cell layer (potential between interdigitated electrodes) (Hille et al. 2009).

4.5.4 Benzimidazol-2-ylidene–Gold(I)

Also, we studied the influence of the benzimidazol-2-ylidene gold(I) complexes **9** and **10** on cellular metabolism and morphological changes of MCF-7 cells in real time using the Bionas 2500 system (Rubbiani et al. 2010) (Fig. 4.20). A rapid reduction of oxygen consumption was observed in all applied concentrations of **9**. In contrast, cellular glycolysis increased at the same time as oxygen consumption was reduced, suggesting an enhanced glycolysis to counteract the inhibition of cellular respiration. However, glycolysis only increased initially and decreased in a dose-dependent manner (Fig. 4.21). Cellular impedance of the cells started to decrease in a concentration-dependent manner approx. 7 h from the onset of treatment with **9** (Fig. 4.21). This indicates changes of cellular morphology, membrane integrity, cell–cell and cell–matrix contacts and an induction and onset of cell death. Interestingly, there was no recovery of cellular metabolism and impedance during the drug-free phase, which indicates irreversible cellular damage. In contrast, compound **10** showed only minor effects on MCF-7 cells at a high concentration (10 μM). This clearly indicates the importance of the chelated central gold(I) atom.

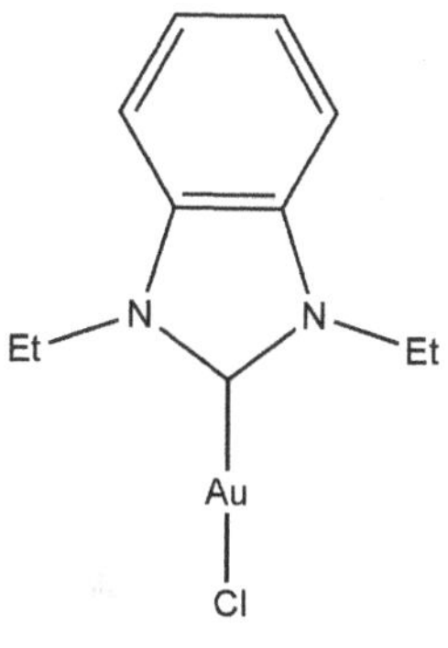

9 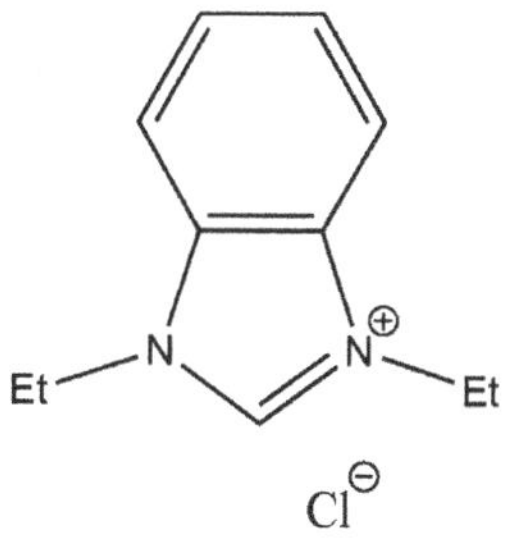   **10**

Fig. 4.20. Benzimidazol-2-ylidene complexes

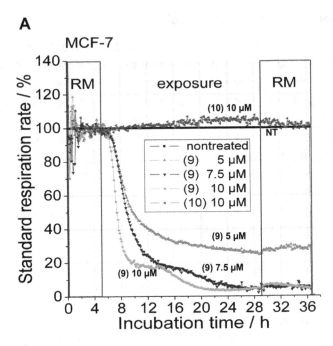

A

MCF-7

RM | exposure | RM

Standard respiration rate / %

140
120
100
80
60
40
20
0

(10) 10 µM

NT

(9) 5 µM

(9) 7.5 µM

(9) 10 µM

- nontreated
- (9)   5 µM
- (9)   7.5 µM
- (9)   10 µM
- (10) 10 µM

0 4 8 12 16 20 24 28 32 36

Incubation time / h

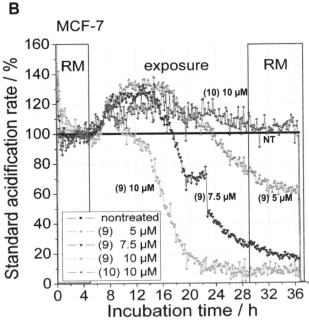

B

MCF-7

RM | exposure | RM

Standard acidification rate / %

160
140
120
100
80
60
40
20
0

(10) 10 µM

NT

(9) 10 µM

(9) 7.5 µM

(9) 5 µM

- nontreated
- (9)   5 µM
- (9)   7.5 µM
- (9)   10 µM
- (10) 10 µM

0 4 8 12 16 20 24 28 32 36

Incubation time / h

90

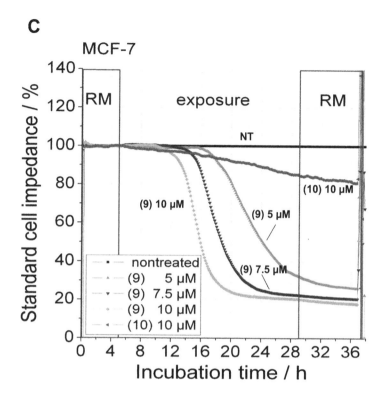

C

MCF-7

Fig. 4.21. Benzimidazol-2-ylidene–gold(I) complex – real-time response profiles of cellular respiration, glycolysis, and impedance of MCF-7. Treatment started after 5 h of equilibration and was continued for 24 h. *RM* (running medium) marks compounds free medium; *exposure* marks the time when compound **9** and **10** were presented at the indicated concentrations (5, 7.5, and 10 μM). **A**, change in respiration (oxygen in the medium); **B**, glycolysis, depicted as change in acidification of the medium (extracellular pH); **C**, change in impedance of the cell layer (potential between interdigitated electrodes) (Rubbiani et al. 2010).

5 Discussion

5.1 Metabolism of Cancer Cells – Differences in the Basic Metabolic Rate

The basic metabolic rate varies between the different cancer cell lines investigated in this study (the two breast cancer cell lines MCF-7 and MDA-MB-231, the two colorectal cancer cell lines HT-29 and HCT-116, and the liver hepatocellular carcinoma HepG2). In particular, there is a pronounced difference in the levels of glycolysis and respiration between MCF-7 (lowest acidification and highest respiration rates) and the other employed cell lines (see Fig. 4.1).

It is clear that this also means that each cell line responds differently towards cytotoxic compounds: the stress-inducing compounds methyl methanesulfonate (MMS) and *tert*-butyl hydroperoxide (*t*BHP) caused MCF-7 to enhance glycolysis as respiration was inhibited, while HT-29 showed an increase of respiration (see Fig. 4.2).

MCF-7 and HT-29 respond differently to rhodium(III) polypyridyl (Harlos et al. 2008): while MCF-7 cells are able to increase glycolysis upon treatment with rhodium complex, HT-29 is not able to do so; in contrast, respiration of HT-29 increased at the beginning of exposure, but not in MCF-7. Also we could show that respiration inhibition by gold complexes in MCF-7 leads to a significant increase of glycolysis (Rubbiani et al. 2010).

The basal rates of respiration and glycolysis of cancer cells need to be taken into consideration in chemotherapy. Improved anticancer treatment may become possible by simultaneously applying drugs that manipulate glycolysis and/or respiration along with other cytotoxic agents.

5.2 Energy Metabolism and Glucose Availability

Cell proliferation in normal cells is controlled by the availability of nutrients, oxygen, and growth factors supplied by the blood circulation. In contrast, in tumors particularly in the initial stage of tumor formation and because of lack of formation of new blood vessels, cancerous cells rely mostly on glycolysis (Warburg effect) as mentioned earlier. The metabolic changes recorded in real-time showed that the metabolic response can be strongly affected by cell environmental conditions. Of course, one of the major players is glucose, as the main source of energy for cells. The amount of glucose in the medium has a pronounced effect on cellular activities, in particular on the rate of respiration and glycolysis. In HT-29 cells the respiration rate decreases for low glucose concentrations, while at the same time the rate of glycolysis increases (at < 0.3 g/L). This phenomenon known as 'Crabtree effect' was observed in tumors where respiration and glycolysis compete for ADP (Wojtczak 1996). We therefore chose 1 g/L of glucose in our standard medium, which is equivalent to the normal blood glucose concentration. However, mostly we used a medium containing 4.5 g/L glucose, which is in excess of the physiological concentration, but since we have constant change of medium in the Bionas system we decided to use 1 g/L. This glucose concentration had no down- or up-regulating effect on respiration or glycolysis (see Fig. 4.3). Glucose concentration also can play a role in cellular responses to cytotoxic compounds and as we will discuss later in this section, cotreatment of anticancer drugs with compounds known as glucose metabolism regulators can change the cytotoxicity of the compounds.

Phlorizin (SGLT glucose transporter inhibitor) caused HT-29 to increase respiration and reduce glycolysis, which can be explained by the lower intracellular glucose concentration. In contrast, phloretin (GLUT glucose transporter inhibitor) caused a reduction of respiration as well as a reduction of glycolysis (after a sharp and immediate increase) (see Fig. 4.4). This probably indicates a higher glucose concentration within cells since GLUT2 is mostly located on the basolateral surfaces of colon cells and is responsible for equilibrating the glucose concentration between enterocyte and plasma (Kellett and Brot-Laroche 2005) (Fig. 5.1).

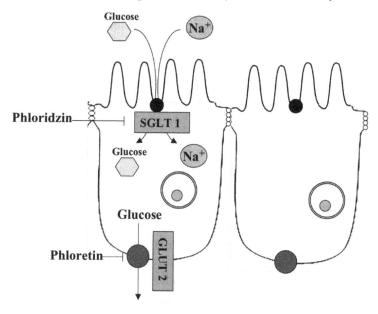

Apical surface (intestinal lumen)

Glucose

Na⁺

Phloridzin

SGLT 1

Glucose

Na⁺

Glucose

Phloretin

GLUT 2

Basolateral surface (blood)

Fig. 5.1. Glucose transport through intestinal epithelia. Phloridzin (phlorizin) inhibits SGLT1 – phloretin inhibits GLUT2

5.3 Regulation of Energy Metabolism by Sirtuin Deacetylase SIRT3

SIRT3 has been overexpressed in HeLa cells and the metabolic changes monitored in real time on the biosensor chips. By these means, the metabolic effects of a single protein can be monitored over an extended period of time (see Fig. 4.5). SIRT3 appears to directly activate the complex I of the respiration chain in mitochondria (Ahn et al. 2008).

For the first time it was possible to obtain a kinetic increase of respiration upon SIRT3 overexpression in real time – this increase amounted to 30–40% as compared to controls (see Fig. 4.5 C). Western blot analysis could confirm the level and timing of SIRT3 overexpression (see Fig. 4.5 B). Both real-time measurement and Western blot showed that the levels of respiration and SIRT3 reached a maximum at ~20 h post transfection (see Fig. 4.5 A). As the transfection efficiency was only about 50–60%, the 30–40% increase of respiration measured in real time needs to be doubled. This is also in good agreement with a recent report that SIRT3 overexpression can increase cellular respiration by 80% (Shi et al. 2005).

A further advantage of real-time measurements is the ability to monitor an overexpressed protein simultaneously with an administered drug – thus providing additional information on the impact of overexpressed proteins upon drug treatment. However, one has to take into account some limitations in respect to the overexpression of a protein, i.e., (i) transfection harms and may reduce the number of viable cells, (ii) the transfection efficiency is critical for the outcome of an experiment. Therefore, the less stressful electroporation for introducing foreign genes into the cells rather than lipofection will result in better cell survival and transfection efficiency and therefore clearer results.

5.4 Cisplatin – Its Effects on Metabolism

5.4.1 Time Points of Metabolic Changes in Cisplatin Treatment

Cisplatin induces changes in cellular metabolism before onset of cell death. In this regard, the first cellular responses observed after initiation of treatment provide specific time-resolved insight into cisplatin's toxicity and hinting to the possible molecular mechanism.

Oxygen consumption is the first parameter affected by cisplatin, either almost immediately after onset of treatment (HT-29, HCT-116, HepG2, and MDA-MB-231) or after 5–6 h (MCF-7) (see Fig. 4.6 A, D, G, J). This appears to confirm the hypothesis that mitochondria are a primary target of cisplatin-mediated DNA damage (Garrido et al. 2008; Cullen et al. 2007).

The latter studies indicate that cisplatin accumulates in mitochondria, binds to mitochondrial DNA (mtDNA), and causes mitochondrial damage. Our results with isolated mitochondria showed no immediate effect on mitochondrial activity within a time frame during which respiration of intact cells is clearly reduced (see Fig. 4.9). This correlates with previous observations that lower concentrations of cisplatin do not inhibit activity of isolated rat liver mitochondria (Garrido et al. 2008). Thus, the immediate respiratory response to cisplatin must result from a mitochondrial activity-independent mechanism involving cellular signaling or DNA damage. However, because it is also observed in the p53 mutant cell line MDA-MB-231, DNA damage-mediated signaling may not be the primary cause.

p53 is a crucial mediator of DNA-damage response signaling. The decrease in respiration in MDA-MB-231 can neither result from DNA-damage sensing nor from direct mitochondrial inhibition. Thus, the decrease in respiration must be mediated by another unknown mechanism. As shown previously, thioredoxin reductase (TrxR) can be a target of electrophilic compounds like cisplatin and thus become inactivated (Witte et al. 2005). Since TrxR plays a key role in the last step of the electron transport chain in mitochondria, a reduction of oxygen consumption in mitochondria should have resulted, but has not been observed in the current trials, however. Thus, it still remains an open question as to why in some cells we observed immediate respiration inhibition, while no immediate respiration change could be observed in isolated mitochondria exposed to cisplatin.

Glycolysis decreased after 6–7 h in HCT-116 and HepG2, at about the same time or slightly before impedance decreased. In striking contrast, glycolysis was stimulated in MCF-7 at ~8 h after onset of treatment, clearly before cells started to die in the biosensor assay. In the cisplatin-resistant cell line MDA-MB-231 no change in glycolysis was observed (see Figs. 4.6 B, E, H, K and 4.7 B).

Impedance measurements on cisplatin-treated MCF-7 showed a rapid decline at 10–11 h in a dose-dependent manner, while HCT-116,

HepG2, and HT-29 display a lower rate of impedance decrease (lower slope) – thus MCF-7 is clearly more sensitive to cisplatin than the other cell lines (see Fig. 4.6 C, F, I, L). The lower sensitivity of HCT-116, HepG2, and HT-29 to cisplatin could possibly be linked to the higher basic levels of glycolysis (see Sect. 5.1).

In a recent study centered on cisplatin and 2-deoxy-D-glucose cotreatment, it was suggested that a high basal glycolytic rate correlates with a more cisplatin-resistant phenotype in two epithelial ovarian carcinomas (Hernlund et al. 2009) – own results seem to confirm this observation. Although the rate of impedance reduction is different, all cell lines showed ongoing cell death when cisplatin was removed after 24 h, indicating irreversible cellular damage. This is in good agreement with DNA damage as being the main mode of cisplatin action (Chaney et al. 2005; Zdraveski et al. 2002; Wang and Lippard 2005; Vogelstein et al. 2000).

5.4.2 Signaling Molecules Akt1, GSK-3ß, and ERK1 Are Affected After Long-Time Exposure to Cisplatin

The activation of Akt1, GSK-3ß, and ERK1/2 MAP-kinases was not significant affected within the first 8–11 h (glycolysis and impedance change) of cisplatin treatment. However, the significant decrease in p-Akt1 and p-GSK-3ß after 24 h reflects a reduced pro-survival signaling, which was more pronounced after a shorter pulse of 10-h treatment with 50 μM cisplatin and 14-h drug-free medium administration than with lower (10 μM) continuous treatment for 24 h (see Fig. 4.11).

Cisplatin treatment has been shown to be associated with specific activation of signaling pathways mediating DNA damage response and cellular proliferation, including p53, MAP-kinases ERK1/2, and Akt1 (Kim et al. 2008; Datta et al. 1999; Panka et al. 2008). ERK1 contributes to apoptosis by activating tumor suppressor p53 (Wang et al. 2000; Kim et al. 2005). The specific efficiency of 50 μM cisplatin treatment is also supported by an increased level of p-ERK1, which also contributes to apoptosis induction (Kim et al. 2005).

To further determine the time points of cisplatin impact in MCF-7, gene expression profiles were analyzed at the particular time points of change of glycolysis and impedance (see Fig. 4.12 and Table 4.2).

5.4.3 Apoptosis Initiation by Cisplatin – Time Point of Onset as Determined by Gene Expression Profiling

This specific pro-apoptotic regulation is also confirmed by changes in gene expression, which include a clear induction of pro-apoptotic genes involved in p53-mediated cell fate decision, such as TSP1, maspin, PMAIP1(NOXA), and GADD45a (Wang and Lippard 2005). Expression of these genes was already induced when glycolysis changed and for some genes increased even further until onset of cell death. In contrast, the stress response genes HSP70, DUSP1, ID2, and CYR61 were strongly regulated at glycolysis change without further induction (Shim et al. 2002).

This suggests that in response to cisplatin, first an initial general stress response is activated before pro-apoptotic cell fate decision occurs. The time frame observed in own online measurements indicates that this crucial transition occurs at ~10 h after onset of cisplatin treatment. Surprisingly, this is in good agreement with a recently published model of p53-dependent cell fate decision in response to DNA damage (Pu et al. 2010).

Our results also show that p53 target genes are involved in this cell fate decision. Furthermore, the primary p53 regulatory circuit is clearly modulated towards p53 activation and pro-apoptotic signaling by E2F1 and E2F3, the one by an up-regulated and the other by a down-regulated mechanism (see Fig. 4.12 C).

In essence, here for the first time a clear time line for cisplatin response has been shown in different cancer cell lines. Also once more it has been confirmed that the p53 regulatory circuit is crucial for the pro-apoptotic cell fate decision, which is further supported by the good agreement of own data with modeling of the p53 damage response (Pu et al. 2010).

Although in our results no significant ROS formation could be observed (Fig. 4.10), the other results clearly show that stress response signaling precedes the pro-apoptotic response (Fig. 5.2).

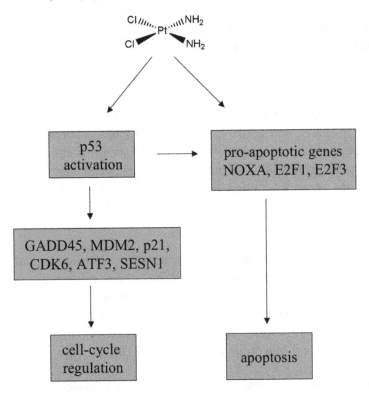

Fig. 5.2. **DNA damage and p53 induction signaling** towards cell cycle arrest and apoptosis (modified after Wang et al. 2009)

5.5 Organometallics – New Anticancer Candidates

5.5.1 Ruthenium(II) Polypyridyls: Impact on Cellular Impedance in HT-29

Continuous profiling of HT-29 response to ruthenium(II) polypyridyl compounds (**1–5**) showed only slight changes of glycolysis, while compound **5** reduced respiration by 40% and cellular impedance by up to 70% after 24 h treatment (see Fig. 4.15).

This strong impact on impedance by **5** and low impact by **1–4** indicates a direct interaction of **5** with the cell membrane and its proteins and thus may effect cell-cell and cell-matrix interactions (Table 5.1).

The molar amount of ruthenium uptake can be estimated based on cell volume and cell protein content – 1 ng/mg cell protein corresponds to 2.0 μM ruthenium compound in HT-29 cells (Table 5.1); the concentration of **5** reached 325 μM in HT-29, which is well above the applied exposure concentration (100 μM) while the uptake by **1–4** was significantly lower, resulting in lower biological potencies of the latter. More efficient cellular accumulation thus leads to higher cytotoxicity.

Further investigations seem warranted on the effect of **5** on cell membranes and cell-cell contacts.

Table 5.1. Cellular uptake and cytotoxicity of ruthenium complexes 1–5 in HT-29 (Schatzschneider et al. 2008).

Compound	IC$_{50}$ (μM)	Ru (ng/mg cell protein)
1	32.3 ± 8.4	8.5 ± 3.3[a]
2	119.4 ± 10.8	15 ± 9.3[a]
3	62.1 ± 3.9	9.9 ± 4.8[a]
4	26.9 ± 0.4	47.8 ± 8.8[a]
5	6.4 ± 1.9	128.2 ± 2.7[b]

5.5.2 Rhodium(III) Polypyridyls: MCF-7 and HT-29 Respond Differently

The rhodium(III) polypyridyl complexes (**6** and **7**) had an immediate effect on MCF-7 and HT-29 – all measured parameters (cellular oxygen consumption, extracellular pH changes, and cellular impedance) responded instantly upon exposure, **6** being the more potent of the two (see Fig 4.17). First, respiration was reduced followed by an increase of glycolysis, but only in MCF-7 cells. This strong effect of **7** on cellular oxygen consumption might be an indication of mitochondria being the primary target of this compound. Also, MCF-7 was more sensitive to both **6** and **7** (see Fig 4.17 D, E, F).

As indicated earlier, the different responses of MCF-7 and HT-29 to these compounds may be due to the striking difference in basic rates of glycolysis and respiration of these two cancer cell lines.

MCF-7 does not recover from respiratory inhibition by **6** and **7** at the applied low concentration (2 μM), while HT-29 showed transient inhibition at both applied concentrations (2 and 5 μM) of **7** and low concentration (2 μM) of **6** (see Fig. 4.17 A, D).

The strong effect on respiration seems to coincide with a study showing that rhodium intercalators can damage mitochondrial DNA (Merino and Barton 2008). Impedance changes showing modifications of cellular adhesion indicate that cell death and apoptosis are most probably contributed to the biological activity of the compounds. Studies on the cellular uptake of this family of compounds revealed that a significant quantity of the agents is accumulated in the lipophilic membrane possibly causing morphological changes, which appear to correlate to the antiproliferative effects triggered by these agents (Table 5.2).

Our findings call for further investigations of the effects of this family of compounds on the respiratory pathway.

Table 5.2. Cellular uptake of rhodium compound (ng/mg cell protein) by MCF-7 and HT-29 exposed to 1 μM of **6** and **7** (Harlos et al. 2009).

Compound	MCF-7	HT-29
6	92.1 ± 1.4	53.6 ± 2
7	43.1 ± 1.9	69.8 ± 18

5.5.3 Phenylenediamine Iron: Response of MCF-7 Is Immediate and Irreversible

MCF-7 responded immediately to salophene iron complex **8** within the first 3 h upon treatment: both respiration and glycolysis during the first hour of exposure, followed by reduction of impedance as an indication of an onset of apoptosis. Cellular damage was irreversible and continued after treatment was stopped (see Fig. 4.19).

102

ROS formation in both MCF-7 and human T lymphocyte Jurkat cells was significantly increased, with Jurkat cells being somewhat more sensitive (Hille et al. 2009) – ROS seem to trigger apoptosis (Simon et al. 2000).

Since this compound had a very strong effect on cellular respiration, further studies on the impact of this compound on mitochondria seem warranted.

5.5.4 Benzimidazol-2-ylidene–Gold(I): Respiration Is Strongly Affected in MCF-7

Respiration was affected by **9** in a dose-dependent manner (Fig. 4.21 A) (Rubbiani et al. 2010). Impairment of respiration rate was followed by an increase of gylcolysis – possibly a compensation for the inhibition of respiration. The decline of impedance after 5–6 h with 10 μM of **9** indicates the time point of induction of apoptosis. Cellular damage due to **9** was irreversible within the first hour of exposure (see Fig. 4.21).

Further studies revealed that one of the main targets of **9** is TrxR. The strong inhibition of this enzyme in fact coincides with the immediate and sharp decline of respiration; also, **9** strongly induces ROS formation – major inducers of apoptosis and necrosis (Rubbiani et al. 2010). Further analysis of the effect of this compound on isolated mouse liver mitochondria showed also an immediate and dose-dependent mitochondrial oxygen consumption inhibition (Fig. 5.3).

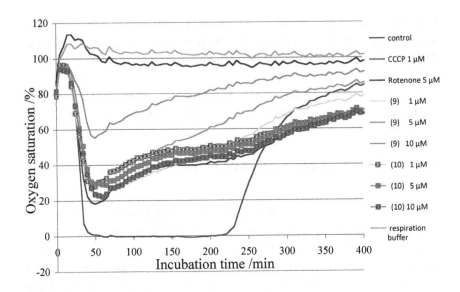

Fig. 5.3. Respiration of freshly isolated mouse liver mitochondria inhibited upon treatment with 9 and 10 in a dose-dependent manner. Controls: rotenone – inhibitor of respiratory chain complex I; and carbonyl cyanide 3-chlorophenyl-hydrazone (CCCP) – decoupler of mitochondrial respiration, leading to increased oxygen consumption (Rubbiani et al. 2010).

6 Conclusion

The cell biosensor chip system not only provides detailed information on cellular metabolism and proliferation but for the first time enabled the precise time-resolved analysis of the cellular response to drug treatment. Analyzing cisplatin as a reference antitumor drug, it could be clearly shown that although a decrease in respiration is the first measurable change, this is not crucial for the cellular response. In contrast, a defined time frame is required from start of treatment until cells start to die, which requires an intact DNA-damage response through p53 signaling. The measurements clearly show that the regulatory steps of p53-mediated cell fate decision are time-dependent and require 10–11 h before cells start to die. This time frame is in very good agreement with recent systems biology modeling of the complex regulatory nodes of p53 signaling.

Another important finding is the interdependence of the basic metabolic activitiy (of untreated tumor cells) and the sensitivity of the cancer cells to cisplatin treatment. Cancer cells with a higher respiratory activity and decreased glycolytic activity seem to be much more sensitive to cisplatin treatment. This suggests that combination of compounds acting on DNA and the DNA-damage response with measurements targeting cellular metabolism should enhance the efficiency of drug treatment.

Besides providing valuable additional information on cisplatin activity and the role of metabolism in antitumor drug therapy, the system can also be used to reveal other so far inaccessible changes in the cell. For example, some new organometallic compounds lead to severe changes on the cell surface without concomitant cell death, the mechanism of which is still not understood. Another important example for the versatility of the system is the online observation of the metabolic change upon overexpression of the key mitochondrial regulator SIRT3, which could be followed over time, clearly demonstrating that fully functional expression occurred.

7 List of Publications

*Gross A, **Alborzinia H**, Piantavigna S, Martin LL, Wölfl S, Metzler-Nolte N (2015) Vesicular disruption of lysosomal targeting organometallic polyarginine bioconjugates. *Metallomics* 7: 371–384

*Cheng X, Holenya P, Can S, **Alborzinia H**, Rubbiani R, Ott I, Wölfl S (2014) A TrxR inhibiting gold(I) NHC complex induces apoptosis through ASK1-p38-MAPK signaling in pancreatic cancer cells. *Mol Cancer* 13: 221

*Holenya P, Can S, Rubbiani R, **Alborzinia H**, Jünger A, Cheng X, Ott I, Wölfl S (2014) Detailed analysis of pro-apoptotic signaling and metabolic adaptation triggered by a N-heterocyclic carbene-gold(I) complex. *Metallomics* 6: 1591–1601

*Kitanovic I, Can S, **Alborzinia H**, Kitanovic A, Pierroz V, Leonidova A, Pinto A, Spingler B, Ferrari S, Molteni R, Steffen A, Metzler-Nolte N, Wölfl S, Gasser G (2014) A deadly organometallic luminescent probe: anticancer activity of a Re I bisquinoline complex. *Chemistry* 20: 2496–2507

*Maschke M, **Alborzinia H**, Lieb M, Wölfl S, Metzler-Nolte N (2014) Structure–activity relationship of trifluoromethyl-containing metallocenes: electrochemistry, lipophilicity, cytotoxicity, and ROS production. *ChemMedChem* 9: 1188-1194

*Meyer A, Oehninger L, Geldmacher Y, **Alborzinia H**, Wölfl S, Sheldrick WS, Ott I (2014) Naphthalimide ligands as combined thioredoxin reductase inhibitors and DNA intercalators. *ChemMedChem* 9: 1794–1800

*Spincemaille P, **Alborzinia H**, Dekervel J, Windmolders P, van Pelt J, Cassiman D, Cheneval O, Craik DJ, Schur J, Ott I, Wölfl S, Cammue BP, Thevissen K (2014) The plant decapeptide OSIP108 can alleviate mitochondrial dysfunction induced by cisplatin in human cells. *Molecules* 19: 15088–15102

***Alborzinia H**, Schmidt-Glenewinkel H, Ilkavets I, Breitkopf K, Cheng X, Hortschansky P, Dooley S, Wölfl S (2013) Quantitative kinetic analysis of BMP2 uptake into cells and its modulation by BMP-antagonists. *J Cell Sci* 126: 117–127

*Oehninger L, Stefanopoulou M, **Alborzinia H**, Schur J, Ludewig S, Namikawa K, Muñoz-Castro A, Köster RW, Baumann K, Wölfl S, Sheldrick WS, Ott I (2013) Evaluation of arene ruthenium(II) N-heterocyclic carbene complexes as organo-metallics interacting with thiol and selenol containing biomolecules. *Dalton Trans* 42:1657–1666

*Cheng X, **Alborzinia H**, Merz KH, Steinbeisser H, Mrowka R, Scholl C, Kitanovic I, Eisenbrand G, Wölfl S (2012) Indirubin derivatives modulate TGFβ/BMP signaling at different levels and trigger ubiquitin mediated depletion of non-activated R-Smads. *Chem Biol* 21:1423–1436

*Geldmacher Y, Splith K, Kitanovic I, **Alborzinia H**, Can S, Rubbiani R, Nazif MA, Wefelmeier P, Prokop A, Ott I, Wölfl S, Neundorf I, Sheldrick WS (2012) Cellular impact and selectivity of half-sandwich organorhodium(III) anticancer complexes and their organoiridium(III) and trichlorido-rhodium(III) counterparts. *J Biol Inorg Chem* 17: 631–646

*Kasper C, **Alborzinia H**, Can S, Kitanovic I, Meyer A, Geldmacher Y, Oleszak M, Ott I, Wölfl S, Sheldrick WS (2012) Synthesis and cellular impact of diene-ruthenium(II) complexes: A new class of organoruthenium anticancer agents. *J Inorg Biochem* 106: 126–133

*Köster SD, **Alborzinia H**, Can S, Kitanovic I, Wölfl S, Rubbiani R, Ott I, Riesterer P, Prokop A, Merz K, Metzler-Nolte N (2012) A spontaneous gold(I)-azide alkyne cycloaddition reaction yields gold-peptide bioconjugates which overcome cisplatin resistance in a p53-mutant cancer cell line. *Chem Sci* 3: 2062–2072

*Meyer A, Bagowski CP, Kokoschka M, Stefanopoulou M, **Alborzinia H**, Can S, Vlecken, DH, Schuhmann W, Sheldrick WS, Wölfl S, Ott I (2012) On the biological properties of alkynyl gold(I) phosphine bioorganometallics. *Angewandte Chemie, Int Ed Engl* 51: 8895-8899

*Nazif MA, Rubbiani R, **Alborzinia H**, Kitanovic I, Wölfl S, Ott I, Sheldrick WS (2012) Cytotoxicity and cellular impact of dinuclear organoiridium DNA intercalators and nucleases with long rigid bridging ligands. *Dalton Trans* 41: 5587–5598

Alborzinia H, Can S, Holenya P, Scholl C, Lederer E, Kitanovic I, Wölfl S (2011) Real-time monitoring of cisplatin-induced cell death. *PLoS ONE* 6, e19714

Bieda R, Kitanovic I, **Alborzinia H**, Meyer A, Ott I, Wölfl S, Sheldrick WS (2011) Antileukaemic activity and biological function of rhodium(III) crown thiaether complexes. *Biometals* 24: 645–561

Geldmacher Y, Kitanovic I, **Alborzinia H**, Bergerhoff K, Rubbiani R, Wefelmeier P, Prokop A, Ott I, Wölfl S, Sheldrick WS (2011) Cellular selectivity and biological impact of cytotoxic rhodium(III) and iridium(III) complexes containing methyl-substituted phenanthroline ligands. *ChemMedChem* 6: 429–439

Hackenberg F, Oehninger L, **Alborzinia H**, Can S, Kitanovic I, Geldmacher Y, Kokoschka M, Wölfl S, Ott I, Sheldrick WS (2011) Highly cytotoxic substitutionally inert rhodium(III) tris(chelate) complexes: DNA binding modes and biological impact on human cancer cells. *J Inorg Biochem* 105: 991–999

Oehninger L, **Alborzinia H**, Ludewig S, Baumann K, Wölfl S, Ott I (2011) From catalysts to bio-active organometallics: do Grubbs catalysts trigger biological effects? *ChemMedChem* 6: 2142–2145

Rubbiani R, Can S, Kitanovic I, **Alborzinia H**, Stefanopoulou M, Kokoschka M, Mönchgesang S, Sheldrick W, Wölfl S, Ott I (2011) Comparative in-vitro evaluation of N-heterocyclic carbene gold(I) complexes of the benzimidazolylidene type. *J Med Chem* 54: 8646–8657

Dransfeld CL, **Alborzinia H**, Wölfl S, Mahlknecht U (2010) SIRT3 SNPs validation in 640 individuals, functional analyses and new insights into SIRT3 stability. *Int J Oncol* 36: 955–960

Dransfeld CL, **Alborzinia H**, Wölfl S, Mahlknecht U (2010) Continuous multiparametric monitoring of cell metabolism in response to transient over-expression of the sirtuin deacetylase SIRT3. *Clin Epigenet* 1: 55–60

Rubbiani R, Kitanovic I, **Alborzinia H**, Can S, Kitanovic A, Onambele LA, Stefanopoulou M, Geldmacher Y, Sheldrick WS, Wolber G, Prokop A, Wölfl S, Ott I (2010) Benzimidazol-2-ylidene gold(I) complexes are thioredoxin reductase inhibitors with multiple antitumor properties. *J Med Chem* 53: 8608–8618

Hille A, Ott I, Kitanovic A, Kitanovic I, **Alborzinia H**, Lederer E, Wölfl S, Metzler-Nolte N, Schäfer S, Sheldrick WS, Bischof C, Schatzschneider U, Gust R (2009) [*N*,*N'*-Bis(salicylidene)-1,2-phenylene-diamine]metal complexes with cell death promoting properties. *J Biol Inorg Chem* 14: 711–725

Harlos M, Ott I, Gust R, **Alborzinia H**, Wölfl S, Kromm A, Sheldrick WS (2008) Synthesis, biological activity, and structure-activity relationships for potent cytotoxic rhodium(III) polypyridyl complexes. *J Med Chem* 51: 3924–3933

Schatzschneider U, Niesel J, Ott I, Gust R, **Alborzinia H**, Wölfl S (2008) Cellular uptake, cytotoxicity, and metabolic profiling of human cancer cells treated with ruthenium(II complexes [Ru(bpy)2(N–N)]Cl2 with N–N=bpy, phen, dpq, dppz, and dppn. *ChemMedChem* 3: 1104–1109

* papers marked by asterisk were published after Feb. 2011, the date of submitting this thesis to the Faculty of Biosciences, Heidelberg University

8 References

Ahn BH, Kim HS, Song S, Lee IH, Liu J, Vassilopoulos A, Deng CX, Finkel T (2008) A role for the mitochondrial deacetylase SIRT3 in regulating energy homeostasis. *Proc Natl Acad Sci USA* 105 (38): 14447–14452

Alberts B, Johnson A, Walter P, Lewis J, Raff M, Roberts K, Orme N (2007) Molecular Biology of the Cell, 5th ed. Garland/Taylor & Francis, New York

Cassano A, Pozzo C, Corsi DC, Fontana T, Noviello MR, Astone A, Barone C (1995) Effect of cisplatin in advanced colorectal cancer resistant to 5-fluorouracil plus (*S*)-leucovorin. *J Cancer Res Clin Oncol* 121(8): 474–477

Chaney SG, Campbell SL, Bassett E, Wu Y (2005) Recognition and processing of cisplatin- and oxaliplatin-DNA adducts. *Crit Rev Oncol Hematol* 53(1): 3–11

Clark Jr LC, Wolf R, Granger D, Taylor Z (1953) Continuous recording of blood oxygen tensions by polarography. *J Appl Physiol* 6(3): 189–193

Cullen KJ, Yang Z, Schumaker L, Guo Z (2007) Mitochondria as a critical target of the chemotheraputic agent cisplatin in head and neck cancer. *J Bioenerg Biomembr* 39(1): 43–50

Datta SR, Brunet A, Greenberg ME (1999) Cellular survival: a play in three Akts. *Genes Dev* 13(22): 2905–2927

Dransfeld CL, Alborzinia H, Wölfl S, Mahlknecht U (2010) Continuous multiparametric monitoring of cell metabolism in response to transient over-expression of the sirtuin deacetylase SIRT3. *Clin Epigenet* 1(1–2): 55–60

Ehret R, Baumann W, Brischwein M, Lehmann M, Henning T, Freund I, Drechsler S, Friedrich U, Hubert ML, Motrescu E, Kob A, Palzer H, Grothe H, Wolf B (2001) Multiparametric microsensor chips for screening applications. *Fresenius J Anal Chem* 369(1): 30–35

Garrido N, Pérez-Martos A, Faro M, Lou-Bonafonte JM, Fernández-Silva P, López-Pérez MJ, Montoya J, Enríquez JA (2008) Cisplatin-mediated impairment of mitochondrial DNA metabolism inversely correlates with glutathione levels. *Biochem J* 414(1): 93–102

Gavin EJ, Song B, Wang Y, Xi Y, Ju J (2008) Reduction of Orc6 expression sensitizes human colon cancer cells to 5-fluorouracil and cisplatin. *PLoS One* 3(12): e4054

Giaever I, Keese CR (1991) Micromotion of mammalian cells measured electrically. *Proc Natl Acad Sci USA* 88(17): 7896–7900

Harris SL, Levine AJ (2005) The p53 pathway: positive and negative feedback loops. *Oncogene* 24(17): 2899–2908

Harlos M, Ott I, Gust R, Alborzinia H, Wölfl S, Kromm A, Sheldrick WS (2008) Synthesis, biological activity, and structure-activity relationships for potent cytotoxic rhodium(III) polypyridyl complexes. *J Med Chem* 51(13): 3924–3933

Hartinger CG, Zorbas-Seifried S, Jakupec MA, Kynast B, Zorbas H, Keppler BK (2006) From bench to bedside – preclinical and early clinical development of the anticancer agent indazolium *trans*-[tetrachlorobis(1H-indazole)ruthenate(III)] (KP1019 or FFC14A). *J Inorg Biochem* 100(5-6): 891–904

Hernlund E, Hjerpe E, Avall-Lundqvist E, Shoshan M (2009) Ovarian carcinoma cells with low levels of beta-F1-ATPase are sensitive to combined platinum and 2-deoxy-D-glucose treatment. *Mol Cancer Ther* 8(7): 1916–1923

Hille A, Ott I, Kitanovic A, Kitanovic I, Alborzinia H, Lederer E, Wölfl S, Metzler-Nolte N, Schäfer S, Sheldrick WS, Bischof C, Schatzschneider U, Gust R (2009) [*N,N'*-Bis(salicylidene)-1,2-phenylenediamine]metal complexes with cell death promoting properties. *J Biol Inorg Chem* 14(5): 711–725

Hirschey MD, Shimazu T, Goetzman E, Jing E, Schwer B, Lombard DB, Grueter CA, Harris C, Biddinger S, Ilkayeva OR, Stevens RD, Li Y, Saha AK, Ruderman NB, Bain JR, Newgard CB, Farese RV Jr, Alt FW, Kahn CR, Verdin E (2010) SIRT3 regulates mitochondrial fatty-acid oxidation by reversible enzyme deacetylation. *Nature* 464(7285): 121–125

Hosack DA, Dennis G Jr, Sherman BT, Lane HC, Lempicki RA (2003) Identifying biological themes within lists of genes with EASE. *Genome Biol* 4(10): R70

Huang DW, Sherman BT, Lempicki RA (2009) Systematic and integrative analysis of large gene lists using DAVID bioinformatics resources. *Nat Protoc* 4(1): 44–57

Jamieson ER, Lippard SJ (1999) Structure, recognition, and processing of cisplatin-DNA adducts. *Chem Rev* 99(9): 2467–2498

Kellett GL, Brot-Laroche E (2005) Apical GLUT2: a major pathway of intestinal sugar absorption. *Diabetes* 54(10): 3056–3062

Kim HS, Hwang JT, Yun H, Chi SG, Lee SJ, Kang I, Yoon KS, Choe WJ, Kim SS, Ha J (2008) Inhibition of AMP-activated protein kinase sensitizes cancer cells to cisplatin-induced apoptosis via hyperinduction of p53. *J Biol Chem* 283(7): 3731–3742

Kim YK, Kim HJ, Kwon CH, Kim JH, Woo JS, Jung JS, Kim JM (2005) Role of ERK activation in cisplatin-induced apoptosis in OK renal epithelial cells. *J Appl Toxicol* 25(5): 374–382

Koberle B, Masters JR, Hartley JA, Wood RD (1999) Defective repair of cisplatin-induced DNA damage caused by reduced XPA protein in testicular germ cell tumours. *Curr Biol* 9(5): 273–276

Koshy N, Quispe D, Shi R, Mansour R, Burton GV (2010) Cisplatin-gemcitabine therapy in metastatic breast cancer: Improved outcome in triple negative breast cancer patients compared to non-triple negative patients. *Breast* 19(3): 246–248

Kroll TC, Wölfl S (2002) Ranking: a closer look on globalisation methods for normalisation of gene expression arrays. *Nucleic Acids Res* 30(11): e50

Lacour S, Hammann A, Grazide S, Lagadic-Gossmann D, Athias A, Sergent O, Laurent G, Gambert P, Solary E, Dimanche-Boitrel MT (2004) Cisplatin-induced CD95 redistribution into membrane lipid rafts of HT29 human colon cancer cells. *Cancer Res* 64(10): 3593–3598

Lange TS, Kim KK, Singh RK, Strongin RM, McCourt CK, Brard L (2008) Iron(III)-salophene: an organometallic compound with selective cytotoxic and anti-proliferative properties in platinum-resistant ovarian cancer cells. *PLoS One* 3(5): e2303

Li C, Wong WH (2001) Model-based analysis of oligonucleotide arrays: expression index computation and outlier detection. *Proc Natl Acad Sci USA* 98(1): 31–6

Marks F, Klingmüller U, Müller-Decker K (2009) Cellular Signal Processing – An Introduction to the Molecular Mechanisms of Signal Transduction. Garland/Taylor & Francis, New York

Marx U, Sandig V (2007) Drug Testing in Vitro – Breakthroughs and Trends in Cell Culture Technology. Wiley-VCH, Weinheim

Merino EJ, Barton JK (2008) DNA oxidation by charge transport in mitochondria. *Biochemistry* 47(6): 1511–1517

Ott I (2009) On the medicinal chemistry of gold complexes as anticancer drugs. *Coord Chem Rev* 253: 1670–1681

Panka DJ, Cho DC, Atkins MB, Mier JW (2008) GSK-3beta inhibition enhances sorafenib-induced apoptosis in melanoma cell lines. *J Biol Chem* 283(2): 726–732

Pegram MD, Pienkowski T, Northfelt DW, Eiermann W, Patel R, Fumoleau P, Quan E, Crown J, Toppmeyer D, Smylie M, Riva A, Blitz S, Press MF, Reese D, Lindsay MA, Slamon DJ (2004) Results of two open-label, multicenter phase II studies of docetaxel, platinum salts, and trastuzumab in HER2-positive advanced breast cancer. *J Natl Cancer Inst* 96(10): 759–769

Pu T, Zhang XP, Liu F, Wang W (2010) Coordination of the nuclear and cyto-plasmic activities of p53 in response to DNA damage. *Biophys J* 99(6): 1696–1705

Rabbani F, Sheinfeld J, Farivar-Mohseni H, Leon A, Rentzepis MJ, Reuter VE, Herr HW, McCaffrey JA, Motzer RJ, Bajorin DF, Bosl GJ (2001) Low-volume nodal metastases detected at retroperitoneal lymphadenectomy for testicular cancer: pattern and prognostic factors for relapse. *J Clin Oncol* 19(7): 2020–2025

Rebillard A, Tekpli X, Meurette O, Sergent O, LeMoigne-Muller G, Vernhet L, Gorria M, Chevanne M, Christmann M, Kaina B, Counillon L, Gulbins E, Lagadic-Gossmann D, Dimanche-Boitrel MT (2007) Cisplatin-induced apoptosis involves membrane fluidification via inhibition of NHE1 in human colon cancer cells. *Cancer Res* 67(16): 7865–7874

Rosenberg B, VanCamp L, Krigas T (1965) Inhibition of cell division in *Escherichia coli* by electrolysis products from a platinum electrode. *Nature* 205: 698–699

Rosenberg B, VanCamp L, Trosko JE, Mansour VH (1969) Platinum compounds: a new class of potent antitumour agents. *Nature* 222(5191): 385–386

Rubbiani R, Kitanovic I, Alborzinia H, Can S, Kitanovic A, Onambele LA, Stefanopoulou M, Geldmacher Y, Sheldrick WS, Wolber G, Prokop A, Wölfl S, Ott I (2010) Benzimidazol-2-ylidene gold(I) complexes are thioredoxin reductase inhibitors with multiple antitumor properties. *J Med Chem* 53(24): 8608–8618

Schatzschneider U, Niesel J, Ott I, Gust R, Alborzinia H, Wölfl S (2008) Cellular uptake, cytotoxicity, and metabolic profiling of human cancer cells treated with ruthenium(II) polypyridyl complexes [Ru(bpy)$_2$(N–N)]Cl$_2$ with N–N=bpy, phen, dpq, dppz, and dppn. *ChemMedChem* 3(7): 1104–1109

Shelley MD, Burgon K, Mason MD (2002) Treatment of testicular germ-cell cancer: a cochrane evidence-based systematic review. *Cancer Treat Rev* 28(5): 237–253

Shi T, Wang F, Stieren E, Tong Q (2005) SIRT3, a mitochondrial sirtuin deacetylase, regulates mitochondrial function and thermogenesis in brown adipocytes. *J Biol Chem* 280(14): 13560–13567

Shim EH, Kim JI, Bang ES, Heo JS, Lee JS, Kim EY, Lee JE, Park WY, Kim SH, Kim HS, Smithies O, Jang JJ, Jin DI, Seo JS (2002) Targeted disruption of hsp70.1 sensitizes to osmotic stress. *EMBO Rep* (9): 857–861

Simon HU, Haj-Yehia A, Levi-Schaffer F (2000) Role of reactive oxygen species (ROS) in apoptosis induction. *Apoptosis* 5(5): 415–418

Tarpey MM, Wink DA, Grisham MB (2004) Methods for detection of reactive metabolites of oxygen and nitrogen: *in vitro* and *in vivo* considerations. *Am J Physiol Regul Integr Comp Physiol* 286(3): R431–444

Vogelstein B, Lane D, Levine AJ (2000) Surfing the p53 network. *Nature* 408(6810): 307–310

Wang D, Lippard SJ (2005) Cellular processing of platinum anticancer drugs. *Nat Rev Drug Discov* 4: 307–320

Wang X, Martindale JL, Holbrook NJ (2000) Requirement for ERK activation in cisplatin-induced apoptosis. *J Biol Chem* 275(50): 39435–39443

Wiltshaw E, Kroner T (1976) Phase II study of *cis*-dichlorodiammineplatinum(II) (NSC-119875) in advanced adenocarcinoma of the ovary. *Cancer Treat Rep* 60(1):55–60

Witte AB, Anestål K, Jerremalm E, Ehrsson H, Arnér ES (2005) Inhibition of thioredoxin reductase but not of glutathione reductase by the major classes of alkylating and platinum-containing anticancer compounds. *Free Radic Biol Med* 39(5):696–703

Wittes RE, Cvitkovic E, Shah J, Gerold FP, Strong EW (1977) *cis*-Dichlorodiammine-platinum(II) in the treatment of epidermoid carcinoma of the head and neck. *Cancer Treat Rep* 61(3):359–366

Wojtczak L (1996)The Crabtree effect: a new look at the old problem. *Acta Biochim Pol* 43(2):361–368

Wölfl S, Dummer A, Pusch L, Pfalz M, Wang L, Clement JH, Leube I, Ebricht R (2005) Analyzing proteins and proteins modifications with ArrayTube™ antibody microarrays. In: Schena MI (ed) Protein Microarrays. Jones and Bartlett, London, pp 159–168

Zamble DB, Jacks T, Lippard SJ (1998) p53-dependent and -independent responses to cisplatin in mouse testicular teratocarcinoma cells. *Proc Natl Acad Sci USA* 95(11):6163–6168

Zdraveski ZZ, Mello JA, Farinelli CK, Essigmann JM, Marinus MG (2002) MutS preferentially recognizes cisplatin- over oxaliplatin-modified DNA. *J Biol Chem* 277(2):1255–1260

Appendix

Supplementary Table. Genes regulated by cisplatin in MCF-7 cells at change of (i) glycolysis (8–9 h) and (ii) impedance (10–11 h); original Excel file with Affymetrix ID, Gene ID, and accession numbers available by request from the author and from Stefan Wölfl, Heidelberg University, Germany.

Gene Name	(i) Fold Change Glycolysis	(ii) Fold Change Impedance	Imp/Glyc
HSPA6	9.61	12.35	1.29
HSPA6	4.91	4.18	0.85
EIF1	4.77	8.90	1.87
FLJ44342	4.24	5.22	1.23
LOC284801	4.22	3.58	0.85
HEXIM1	4.00	5.90	1.47
PMAIP1	4.00	5.79	1.45
Hs,155364,0	3.81	4.51	1.18
ATF3	3.14	4.14	1.32
FLJ44342	3.11	5.01	1.61
ETS2	3.05	4.09	1.34
GPR27	2.97	2.46	0.83
RGS2	2.93	3.74	1.28
SERTAD1	2.92	3.04	1.04
PHLDA1	2.87	3.83	1.33
PMAIP1	2.86	3.02	1.06
GADD45A	2.78	3.04	1.10
Hs,41272,0	2.74	3.29	1.20
C1orf63	2.69	3.19	1.19
Hs,293560,0	2.66	2.43	0.91
Hs,102981,0	2.65	4.30	1.62
Hs,148497,0	2.65	3.62	1.37
Hs,29692,0	2.65	3.77	1.42
Hs,26507,0	2.63	3.78	1.43
Hs,129037,0	2.62	2.57	0.98
REV3L (yeast)	2.61	3.18	1.22
Hs,132882,0	2.60	3.70	1.42
PADI2	2.56	3.35	1.31
HBA1/2	2.53	3.92	1.55

HBEGF	2.50	3.78	1.51
PSME4	2.50	2.84	1.14
HSPA2	2.49	2.36	0.95
C17orf91	2.49	4.12	1.66
POP4	2.48	3.71	1.50
DDX5	2.46	3.24	1.32
ZNF277	2.46	6.07	2.47
TRIM13	2.45	2.85	1.16
Hs,14613,0	2.43	4.00	1.65
NEAT1	2.41	2.49	1.04
Hs,188809,0	2.35	2.53	1.08
HOXB9	2.35	2.79	1.19
ID2 /// ID2B	2.34	5.00	2.14
ADM	2.33	2.92	1.25
MRPL35	2.32	3.29	1.42
Hs,121525,0	2.32	2.91	1.25
HBA1/2	2.31	4.07	1.76
ZBTB1	2.30	2.72	1.18
Hs,135706,0	2.28	2.46	1.08
CDKN1C	2.28	2.52	1.11
MUC5B	2.28	1.97	0.86
SLC30A1	2.28	2.49	1.09
TNFSF9	2.27	2.74	1.21
Hs,28170,0	2.26	1.65	0.73
Hs,182723,0	2.24	3.09	1.38
ALAS2	2.20	3.04	1.38
TUBB2A	2.19	2.51	1.15
Hs,104580,0	2.19	2.63	1.20
THBS1	2.19	3.24	1.48
IER5L	2.18	2.12	0.97
Hs,292931,0	2.18	3.46	1.59
MSX1	2.17	2.73	1.25
RASD1	2.16	3.97	1.83
NANOS1	2.16	3.17	1.47
RPS11	2.16	1.76	0.82
NUTF2	2.16	3.10	1.44
GUCA1B	2.15	3.44	1.60

FAM109A	2.13	2.01	0.94
CYR61	2.13	2.41	1.13
DUSP1	2.11	2.12	1.01
CDKN1C	2.11	2.68	1.27
H3F3B	2.11	2.43	1.15
Hs,161321,0	2.09	2.37	1.13
Hs,264606,0	2.09	2.53	1.21
Hs,71245,0	2.08	3.14	1.51
Hs,49031,0	2.07	2.83	1.36
ID2	2.07	1.92	0.93
IRF7	2.07	2.83	1.37
HBA1/2	2.06	5.27	2.55
FAM84B	2.06	2.01	0.98
Hs,119940,0	2.06	2.94	1.43
Hs,153944,0	2.06	2.85	1.38
CDKN1C	2.06	2.15	1.05
LOC387763	2.06	3.42	1.66
FAM43A	2.05	2.32	1.13
C7orf40	2.05	2.66	1.30
COX19	2.05	2.78	1.36
Hs,120784,1	2.04	2.19	1.07
MAPK12	2.04	6.07	2.97
Hs2,116691,1	2.04	2.55	1.25
Hs,104349,0	2.04	2.24	1.10
GGT1	2.03	2.63	1.30
Hs,171397,0	2.02	2.71	1.34
HOXC13	2.01	2.34	1.16
HBA1/2	2.01	4.15	2.06
NR4A1	2.01	3.16	1.57
LEAP2	2.01	3.11	1.55
Hs,83938,0	1.99	2.49	1.25
Hs,254477,0	1.99	2.16	1.09
RSL1D1	1.98	2.91	1.47
Hs,105621,0	1.98	2.71	1.37
PDCD6	1.97	2.10	1.06
Hs,6694,0	1.95	3.15	1.61
FOS	1.95	2.26	1.16

CTGF	1.95	2.41	1.24
Hs,127345,0	1.94	2.50	1.29
Hs,160999,1	1.94	2.07	1.07
ERN1	1.93	2.24	1.16
TMED4	1.93	2.15	1.11
WIBG	1.93	2.18	1.13
Hs,178393,0	1.93	2.95	1.53
NRARP	1.92	1.98	1.03
PMS1	1.92	2.07	1.08
HBA1/2	1.92	6.36	3.31
PROKR2	1.92	2.54	1.33
ZNF785	1.91	2.54	1.33
FILIP1L	1.91	1.79	0.93
SERPINB5:	1.91	3.20	1.68
NLGN2	1.91	3.81	1.99
PMS1	1.90	2.16	1.14
CACNG5	1.89	2.70	1.43
C17orf86	1.88	2.30	1.22
HOXA5	1.88	1.74	0.93
DNAJB4	1.86	1.98	1.07
SGK1	1.86	2.42	1.30
RPL37	1.85	2.01	1.09
ARGLU1	1.84	1.80	0.98
C11orf61	1.84	2.19	1.19
LINS1	1.83	1.97	1.08
OVOL1	1.83	2.46	1.34
Hs,144923,0	1.83	1.81	0.99
KLHDC5	1.83	2.33	1.28
HHLA3	1.82	2.31	1.27
ANKRD37	1.82	2.89	1.59
MALAT1	1.82	2.37	1.30
IRGQ	1.82	2.64	1.45
SAP18	1.82	2.21	1.22
USP46	1.82	1.81	1.00
C1orf55	1.81	2.27	1.25
H2AFJ	1.81	2.24	1.24
LHX6	1.80	2.31	1.28

Hs,173422,1	1.80	2.60	1.44
LOC284408	1.80	2.38	1.32
SP6	1.79	2.38	1.33
Hs,59203,0	1.79	2.06	1.15
LOC153546	1.79	2.07	1.15
SNRNP48	1.78	2.25	1.26
ATXN7L2	1.78	2.53	1.42
SRRT	1.78	2.51	1.41
UBR4	1.78	2.14	1.20
C5AR1	1.77	2.34	1.32
Hs2,200745,1	1.77	2.17	1.23
Hs,300496,5	1.77	2.85	1.61
C2orf18	1.77	1.79	1.01
LOC100132288 /// MAFIP	1.76	1.93	1.09
ZNF579	1.75	2.23	1.27
ARL17 /// ARL17P1	1.75	2.80	1.60
TMEM52	1.75	2.24	1.28
Hs,13500,0	1.75	2.69	1.54
PRKAR1B	1.74	2.30	1.32
SNRPE	1.73	1.92	1.11
LMO2	1.73	2.42	1.40
HHLA3	1.73	2.39	1.38
Hs,127121,0	1.73	2.33	1.35
Hs,116932,0	1.73	2.44	1.42
UFM1	1.72	2.29	1.33
ZNF574	1.72	2.07	1.20
Hs,44656,0	1.72	2.14	1.24
CYP2U1	1.72	2.69	1.56
Hs2,123560,1	1.72	2.23	1.29
PSMC3IP	1.72	2.06	1.20
CDKN1C	1.71	2.44	1.42
LOC100290882	1.71	2.63	1.54
Hs,60596,0	1.71	2.18	1.28
RER1	1.70	2.18	1.28
SFRS6	1.70	2.02	1.19
NDUFAF4	1.70	3.46	2.03
ULBP2	1.70	2.11	1.24

HOXC9	1.69	2.06	1.21
Hs,172674,4	1.69	2.00	1.18
RHOB	1.69	2.26	1.34
Hs,268231,0	1.69	2.31	1.37
CD59	1.69	2.32	1.38
POL3S	1.68	2.56	1.52
GPX8	1.68	1.93	1.15
E2F1	1.68	2.54	1.51
ANKRD10	1.68	2.41	1.43
Hs,136316,0	1.68	2.30	1.37
CIRBP	1.68	2.56	1.52
Hs,145569,0	1.67	2.45	1.46
HBA1/2	1.67	4.04	2.42
Hs,132370,4	1.67	1.86	1.11
DPH3	1.67	1.82	1.09
THAP10	1.66	2.01	1.21
NR4A1	1.66	2.25	1.35
Hs,149312,0	1.66	2.02	1.22
TRIAP1	1.66	2.09	1.26
LMTK3	1.66	2.45	1.48
RPL23	1.65	3.03	1.83
Hs,153279,0	1.65	2.51	1.52
C12orf73	1.65	2.16	1.31
CLPB	1.65	2.57	1.56
IER5L	1.65	1.76	1.07
C1orf93	1.65	2.34	1.42
COL1A1	1.65	1.99	1.21
Hs,118947,0	1.64	2.23	1.36
RPS15A	1.64	2.25	1.38
HES6	1.63	2.23	1.36
PIGZ	1.63	1.86	1.14
OGFRL1	1.63	2.46	1.51
SORBS3	1.62	2.15	1.33
ZNF561	1.62	2.06	1.27
SEPP1	1.62	2.49	1.54
LOC100128653	1.62	2.19	1.35
BTG2	1.62	2.01	1.24

RBM8A	1.62	2.16	1.34
ZNF449	1.62	2.09	1.29
Hs,120850,0	1.61	2.23	1.38
C15orf17	1.61	2.23	1.39
TNK1	1.61	1.90	1.18
FOXC1	1.60	2.41	1.50
CYTH2	1.60	2.51	1.57
KIAA1683	1.60	2.81	1.76
RHBDD3	1.60	2.25	1.41
Hs,175569,0	1.60	2.53	1.59
AFFX-r2-Bs-thr-3	1.59	2.84	1.78
C3orf38	1.59	2.26	1.42
DPH2	1.59	1.87	1.18
JUNB	1.59	2.27	1.43
MBLAC1	1.58	2.02	1.28
TMEM183A	1.58	2.14	1.35
KDSR	1.58	1.99	1.27
FGFBP3	1.58	1.99	1.26
CLDN15	1.57	1.81	1.15
DHFRL1	1.57	1.82	1.16
JMJD7-PLA2G4B	1.57	2.12	1.35
Hs,127279,0	1.56	1.92	1.22
MRS2	1.56	2.16	1.38
RCBTB2	1.56	2.03	1.30
LOC100131860	1.56	2.18	1.40
MIB2	1.55	2.40	1.54
LRDD	1.55	2.67	1.72
MTSS1L	1.55	2.64	1.70
OTUB1	1.55	1.99	1.29
FLJ30375	1.55	1.97	1.27
RBM3	1.55	2.29	1.48
PKM2	1.55	1.84	1.19
Hs,12489,0	1.55	2.00	1.29
PDE4A	1.55	2.25	1.46
ZNF263	1.54	1.87	1.21
MAFB	1.53	1.83	1.19
MAFF	1.53	1.97	1.29

DNM1	1.53	2.02	1.32
ZNF750	1.53	2.25	1.47
Hs,29008,0	1.53	2.01	1.32
Hs,14691,0	1.53	1.79	1.17
TXNL4B	1.53	1.84	1.21
EGR1	1.53	1.89	1.24
IP6K2	1.53	1.99	1.30
C1orf163	1.53	1.82	1.19
LOC439911	1.52	1.89	1.24
C1orf109	1.52	1.96	1.29
IGFBP4	1.52	1.77	1.17
C1orf216	1.52	1.97	1.30
SNORA72	1.51	1.83	1.21
ATMIN	1.50	1.95	1.30
CNBP	1.49	2.20	1.47
RBM34	1.49	1.96	1.32
MARCH9	1.48	1.98	1.33
Hs,102941,0	1.47	2.71	1.84
CHAC1	1.47	2.11	1.43
C1orf51	1.47	1.87	1.27
APLP1	1.47	2.17	1.48
ZNF187	1.47	1.88	1.28
PCBD2	1.47	1.97	1.35
EXTL2	1.46	1.94	1.33
TIMM22	1.46	1.98	1.35
SPTAN1	1.46	2.46	1.68
CCNL1	1.46	1.82	1.25
Hs,306793,0	1.46	2.25	1.54
EIF3B	1.46	1.94	1.33
DLG1	1.45	1.84	1.27
C1orf69	1.45	2.16	1.49
SFRS2B	1.45	2.18	1.51
ZNF510	1.44	1.77	1.23
DGCR8	1.44	1.88	1.31
TSSK3	1.43	2.52	1.76
CARD8	1.43	1.80	1.26
LMNA	1.42	1.92	1.35

Hs,51820,0	1.42	1.96	1.37
APOE	1.42	2.13	1.51
TADA2B	1.42	2.05	1.45
LRCH4	1.41	2.35	1.66
EPHX1	1.40	2.38	1.70
TRGV5	1.40	2.74	1.96
SNF8	1.39	1.99	1.43
FSTL3	1.39	2.09	1.50
IGL@ /// IGLV1-44	1.39	2.46	1.78
Hs,24567,0	1.37	1.96	1.43
Hs,186832,0	1.37	2.59	1.89
LOC150166	1.36	1.76	1.30
DTX3L	1.36	1.87	1.38
KCNK12	1.36	1.94	1.43
TRIM21	1.36	1.83	1.35
Hs,298188,0	1.35	1.84	1.36
TSPYL2	1.35	2.08	1.54
C7orf43	1.35	2.00	1.49
ZNF580	1.34	2.43	1.81
C1orf133	1.34	1.79	1.33
CCNJ	1.33	1.88	1.41
HSPA13	1.33	1.77	1.33
THUMPD2	1.33	1.78	1.34
GSTM5	1.33	1.91	1.44
RAB39B	1.31	1.95	1.48
GNG13	1.31	2.32	1.77
Hs,300496,4	1.30	2.05	1.58
PERP	1.29	1.97	1.53
GRINA	1.28	1.85	1.44
MAFF	1.28	2.08	1.63
C20orf160	1.25	1.95	1.55
CCL28	1.25	2.21	1.77
FLYWCH1	1.24	1.80	1.45
IRF3	1.24	1.81	1.46
SIN3B	1.23	1.88	1.53
Hs,36676,0	1.23	2.89	2.34
Hs,209602,0	1.23	2.76	2.24

CHKB-CPT1B	1.22	2.24	1.83
MPPE1	1.21	1.92	1.58
EXOC7	1.19	1.84	1.54
FAM100A	1.18	1.97	1.66
DSCR8	1.15	2.23	1.94
LOC100128309	1.02	2.22	2.18
PCGF1	-1.01	2.89	-2.86
ITPRIPL1	-1.12	3.49	-3.12
JAK3	-1.16	1.96	-1.68
PLEKHA7	-1.28	-1.82	1.42
Hs,247781,0	-1.28	-4.12	3.21
Hs,123307,0	-1.29	3.89	-3.02
Hs,210761,0	-1.30	-1.93	1.48
MGA	-1.33	-1.81	1.36
ATF6	-1.33	-2.23	1.68
ASTN2	-1.34	-1.80	1.34
MAPKAP1	-1.35	-2.09	1.54
MSI2	-1.35	-2.18	1.61
C1orf56	-1.36	-1.95	1.44
NME7	-1.36	-2.12	1.56
SKAP	-1.37	-1.83	1.33
NBEA	-1.38	-2.23	1.62
EIF3B	-1.38	-1.90	1.38
ARMC9	-1.38	-1.96	1.42
GALNT10	-1.39	-2.15	1.55
LRP8	-1.40	-2.13	1.52
SPAG9	-1.41	-1.80	1.28
CUGBP1	-1.41	-1.82	1.29
ATF6	-1.42	-1.99	1.41
GPC6	-1.43	-1.94	1.35
ZFHX3	-1.44	-1.94	1.35
NCAPH	-1.45	-1.87	1.29
KIAA1324L	-1.45	-1.79	1.23
TPD52	-1.46	-2.06	1.42
MAN1A2	-1.46	-1.92	1.32
Hs,283742,0	-1.46	-2.16	1.48
GALNT2	-1.47	-1.80	1.23

NCOA1	-1.47	-2.03	1.39
ZAK	-1.47	-2.27	1.54
ZNF253	-1.47	-2.32	1.57
Hs,122684,1	-1.47	-2.13	1.45
IFT74	-1.48	-2.00	1.36
NUBPL	-1.48	-1.99	1.34
Hs,122684,0	-1.48	-1.97	1.33
CBX7	-1.48	-1.87	1.26
NUMB	-1.48	-1.93	1.30
DDX59	-1.49	-2.06	1.38
RHBDD1	-1.49	-2.13	1.43
ZFAND3	-1.51	-2.18	1.45
MAN1A1	-1.51	-2.06	1.37
MFN1	-1.51	-2.10	1.39
PDS5A	-1.51	-1.86	1.23
CCNY	-1.51	-2.32	1.53
UPRT	-1.51	-1.87	1.24
ATP6V1H	-1.52	-2.15	1.42
NOTCH2NL	-1.52	-1.87	1.23
SMARCA2	-1.52	-1.86	1.22
DENND4C	-1.52	-2.11	1.39
CDCA3	-1.52	-2.28	1.50
MAP4K5	-1.52	-2.14	1.41
C6orf162	-1.52	-2.13	1.40
TLK1	-1.53	-2.23	1.46
RPS6KA3	-1.53	-2.22	1.46
NSMCE2	-1.53	-1.84	1.20
Hs,194477,1	-1.53	-1.80	1.17
NIPBL	-1.54	-1.95	1.27
CKAP2L	-1.54	-1.83	1.19
TMTC1	-1.54	-2.30	1.50
ATG10	-1.54	-2.77	1.80
NDE1	-1.54	-1.93	1.25
Hs,128330,0	-1.54	-2.38	1.54
GULP1	-1.54	-2.21	1.43
TMEM164	-1.55	-1.84	1.19
DLG3	-1.55	-2.00	1.29

CENPC1	-1.55	-1.81	1.17
ANKHD1	-1.55	-1.82	1.18
TDRD3	-1.55	-1.92	1.24
KYNU	-1.55	-2.34	1.51
CUGBP1	-1.56	-2.58	1.66
RAB1A	-1.56	-3.45	2.21
PPFIA1	-1.56	-2.35	1.51
CFLAR	-1.56	-2.08	1.33
MKLN1	-1.56	-2.02	1.29
RDX	-1.56	-1.76	1.13
NFIB	-1.56	-2.04	1.31
PAFAH1B1	-1.56	-2.21	1.42
TLE1	-1.57	-1.92	1.22
RASA1	-1.57	-1.84	1.17
FAM54A	-1.57	-2.58	1.64
MLLT1	-1.58	-1.82	1.15
CCM2	-1.58	-1.97	1.25
SFPQ	-1.58	-1.93	1.22
CDC25C	-1.58	-2.07	1.31
ASCC1	-1.59	-2.05	1.29
KIAA1429	-1.59	-1.82	1.15
TDP1	-1.59	-1.82	1.15
Hs,40289,1	-1.59	-2.73	1.72
C8orf83	-1.60	-2.10	1.32
Hs,114772,0	-1.60	-2.14	1.34
ATF6	-1.60	-3.62	2.26
IREB2	-1.60	-1.83	1.15
RAD18	-1.60	-1.90	1.18
CTNND2	-1.60	-2.20	1.37
RAB2A	-1.61	-2.18	1.35
ASXL2	-1.61	-2.17	1.35
NR3C1	-1.61	-1.97	1.22
SKAP2	-1.61	-3.54	2.20
MSI2	-1.61	-2.13	1.32
UBAC2	-1.61	-2.12	1.31
TBC1D16	-1.62	-1.85	1.14
MAPK14	-1.62	-2.08	1.29

MFSD6	-1.62	-1.88	1.16
CDKAL1	-1.62	-2.47	1.53
ZBTB10	-1.62	-1.95	1.20
BARD1	-1.62	-1.96	1.21
WDR67	-1.62	-2.24	1.38
FAM168A	-1.63	-2.13	1.31
ANKRD27	-1.63	-1.81	1.11
DNMBP	-1.63	-2.55	1.56
RAD18	-1.63	-1.81	1.11
RAPH1	-1.63	-1.89	1.16
DUSP4	-1.64	-1.79	1.09
EPB41L4B	-1.64	-2.19	1.34
COG5	-1.64	-1.86	1.14
LASS6	-1.64	-2.15	1.31
VPRBP	-1.64	-2.00	1.22
COX10	-1.64	-2.22	1.35
PHACTR2	-1.64	-2.63	1.60
CUGBP1	-1.64	-2.11	1.28
Hs,102406,0	-1.65	-2.08	1.27
ARMC8	-1.65	-2.06	1.25
Hs,90035,0	-1.65	-2.08	1.26
DHRS2	-1.65	-3.04	1.84
CADM1	-1.65	-2.02	1.22
SYNJ2	-1.66	-2.25	1.36
MYO1B	-1.66	-1.91	1.15
FAM134B	-1.66	-2.86	1.72
QKI	-1.66	-2.44	1.47
RNF24	-1.67	-2.98	1.79
ERBB2IP	-1.67	-1.95	1.17
SAPS3	-1.67	-1.94	1.16
EHD4	-1.67	-1.93	1.15
WIPF2	-1.67	-1.93	1.16
DTNA	-1.67	-2.14	1.28
THADA	-1.67	-2.03	1.21
PAK2	-1.67	-1.89	1.13
SNTB2	-1.67	-2.40	1.43
NR3C1	-1.67	-2.04	1.22

CDK8	-1.67	-2.00	1.20
WDR26	-1.68	-1.92	1.15
LPCAT1	-1.68	-2.13	1.27
LTBP1	-1.68	-2.20	1.32
KLF7	-1.68	-1.95	1.16
DLG1	-1.68	-2.15	1.28
CHD9	-1.68	-2.51	1.49
Hs,40289,0	-1.68	-1.93	1.15
CHD9	-1.68	-2.22	1.32
CCNY	-1.68	-1.93	1.14
SNX24	-1.69	-2.01	1.19
CAB39	-1.69	-2.06	1.22
ZMYM5	-1.69	-2.46	1.45
SLC6A6	-1.69	-2.14	1.26
CRIM1	-1.69	-2.00	1.18
FOXO3	-1.69	-2.76	1.63
PPP3CA	-1.70	-2.05	1.21
SCYL3	-1.70	-1.95	1.15
ANKRD12	-1.70	-2.64	1.56
INSIG1	-1.70	-2.15	1.26
SENP5	-1.70	-1.82	1.07
AGPAT3	-1.70	-2.09	1.22
AMMECR1	-1.70	-1.90	1.11
MSH2	-1.71	-2.60	1.52
WWC1	-1.71	-2.45	1.43
ARNT	-1.71	-1.86	1.09
GTF2F2	-1.71	-2.43	1.42
CDK8	-1.71	-2.58	1.51
FAM168A	-1.71	-1.95	1.14
SPIRE1	-1.71	-2.04	1.19
MARK3	-1.71	-1.91	1.12
NEBL	-1.71	-2.17	1.27
NFIB	-1.71	-1.99	1.16
SH3D19	-1.71	-2.10	1.22
ATL2	-1.72	-2.17	1.26
PHACTR4	-1.72	-2.36	1.37
SH3BP5	-1.72	-2.49	1.45

MBNL1	-1.72	-2.42	1.41
KYNU	-1.72	-2.96	1.72
DENND2C	-1.73	-2.03	1.18
PDSS2	-1.73	-2.21	1.28
C1orf77	-1.73	-3.32	1.92
PLD1	-1.73	-1.97	1.14
DHRSX	-1.73	-2.86	1.66
IGF1R	-1.73	-3.71	2.14
PJA2	-1.73	-2.53	1.46
IPO11	-1.73	-2.57	1.49
GPBP1	-1.73	-2.14	1.24
KRIT1	-1.73	-2.92	1.69
AP4S1	-1.73	-2.51	1.45
KCNK1	-1.74	-1.75	1.01
BCR	-1.74	-2.08	1.20
MED27	-1.74	-2.22	1.28
UXS1	-1.74	-2.29	1.32
PAK2	-1.74	-2.50	1.44
FBXO32	-1.74	-2.05	1.18
PPAP2A	-1.74	-2.12	1.22
Hs,272458,2	-1.75	-3.03	1.73
KLF3	-1.75	-2.12	1.21
MED6	-1.75	-1.94	1.11
EPS15	-1.75	-2.23	1.27
AP4S1	-1.75	-2.89	1.65
Hs,47083,0	-1.75	-1.85	1.05
KIF2A	-1.76	-1.84	1.05
PARD3	-1.76	-2.45	1.39
RGS17	-1.76	-2.38	1.35
RHBDD1	-1.76	-2.39	1.35
BICD2	-1.76	-2.37	1.34
HTR7P	-1.77	-2.15	1.22
WASL	-1.77	-2.00	1.13
HDHD1A	-1.77	-1.88	1.06
RNF150	-1.77	-2.63	1.48
PREX1	-1.77	-2.62	1.48
ASH1L	-1.77	-2.08	1.17

CUL3	-1.78	-1.92	1.08
C17orf63	-1.78	-2.03	1.15
PUM2	-1.78	-1.94	1.09
EXOSC3	-1.78	-2.69	1.51
SETD2	-1.78	-2.41	1.35
ROCK1	-1.78	-2.77	1.56
USP37	-1.78	-1.88	1.06
GABPB1	-1.78	-2.05	1.15
FOXO3	-1.79	-3.12	1.75
UBE2H	-1.79	-2.00	1.12
KIAA1432	-1.79	-2.62	1.46
CLASP2	-1.79	-1.88	1.05
CLIP1	-1.79	-1.96	1.09
GAB1	-1.80	-1.99	1.11
YAP1	-1.80	-1.93	1.07
ZNF148	-1.80	-2.29	1.27
CRISPLD2	-1.80	-2.38	1.32
PANX1	-1.81	-2.30	1.27
IQSEC1	-1.81	-2.01	1.11
MAPKAP1	-1.81	-3.00	1.66
BBX	-1.81	-2.39	1.32
SRGAP2P1	-1.81	-3.27	1.80
MCM9	-1.81	-1.98	1.09
RSRC1	-1.81	-2.86	1.57
RAB27B	-1.82	-2.29	1.26
PHACTR2	-1.82	-2.60	1.43
C5orf41	-1.82	-2.97	1.63
TAOK3	-1.82	-2.40	1.32
PARD3	-1.82	-2.56	1.41
ZNF292	-1.82	-2.22	1.22
ANKRD13C	-1.82	-2.31	1.27
SP4	-1.82	-2.37	1.30
SPTLC2	-1.83	-2.01	1.10
UBTD2	-1.83	-1.79	0.98
KIAA1217	-1.83	-2.19	1.20
FBXL17	-1.83	-2.08	1.14
RBM27	-1.83	-2.02	1.11

PPP2R5E	-1.83	-2.49	1.36
NFIB	-1.83	-4.43	2.42
ASXL1	-1.83	-1.82	0.99
GAPVD1	-1.83	-1.72	0.94
SLC25A13	-1.83	-2.35	1.28
MYO10	-1.83	-2.33	1.27
DCUN1D1	-1.83	-1.89	1.03
ABCG1	-1.83	-1.86	1.01
MAP2K4	-1.83	-2.39	1.31
BRIP1	-1.83	-2.43	1.33
LARP4B	-1.84	-2.35	1.28
RFWD2	-1.84	-3.94	2.14
ATP2B1	-1.84	-2.08	1.13
SASH1	-1.84	-1.98	1.08
CUX1	-1.84	-2.43	1.32
ASH1L	-1.84	-2.63	1.43
SAFB	-1.84	-3.14	1.70
MBOAT1	-1.84	-1.86	1.01
ATXN7	-1.84	-2.41	1.31
CYTSA	-1.84	-1.83	0.99
ZNF586	-1.84	-1.82	0.99
SEL1L	-1.84	-3.79	2.06
WIPF1	-1.84	-2.22	1.20
SDCCAG10	-1.85	-2.46	1.33
UCK2	-1.85	-2.34	1.27
FUS	-1.85	-2.06	1.12
GLCCI1	-1.85	-2.06	1.11
NCOA3	-1.85	-2.28	1.23
PPP2R5E	-1.85	-2.56	1.39
VPS54	-1.85	-2.13	1.15
SGCG	-1.85	-1.84	1.00
ZCCHC2	-1.85	-2.25	1.21
CUL1	-1.85	-2.17	1.17
MAP3K3	-1.85	-1.71	0.92
SLC2A13	-1.86	-2.99	1.61
NFATC3	-1.86	-2.88	1.55
SLMAP	-1.86	-2.22	1.19

SAPS3	-1.86	-2.38	1.28
B3GNT5	-1.86	-1.62	0.87
STRN3	-1.86	-1.97	1.06
CASK	-1.86	-2.62	1.41
PIAS1	-1.86	-2.15	1.15
RAPGEF1	-1.86	-2.63	1.41
AGPAT3	-1.86	-2.70	1.45
ANKRD13C	-1.86	-1.91	1.02
MRPS31	-1.87	-2.57	1.38
C19orf42	-1.87	-2.32	1.24
DEPDC6	-1.87	-2.11	1.13
MED13L	-1.87	-3.00	1.60
ZNF277	-1.87	-3.15	1.68
WDR20	-1.87	-2.05	1.10
LRRC8D	-1.87	-2.15	1.15
ADAM10	-1.87	-2.91	1.55
JDP2	-1.87	-1.47	0.78
TCF20	-1.87	-1.79	0.95
ZFP36L2	-1.88	-2.57	1.37
DUSP16	-1.88	-2.09	1.11
TRIM24	-1.88	-1.89	1.00
DOCK7	-1.88	-2.18	1.16
SRGAP2	-1.88	-2.35	1.25
PDE7A	-1.88	-1.54	0.82
ZSWIM5	-1.88	-2.15	1.14
RNF13	-1.88	-2.12	1.13
ZMYND8	-1.88	-2.33	1.24
MLL3	-1.88	-2.51	1.33
AGPAT3	-1.89	-2.15	1.14
HMGCLL1	-1.89	-2.02	1.07
VANGL1	-1.89	-2.18	1.15
DIP2C	-1.89	-2.69	1.42
PCDH10	-1.89	-1.89	1.00
ERC1	-1.89	-2.30	1.21
SLC37A3	-1.90	-2.03	1.07
UBE2G1	-1.90	-2.54	1.34
SLC25A24	-1.90	-1.49	0.79

HS2ST1	-1.90	-2.21	1.16
ACVR2A	-1.90	-1.52	0.80
FRMD6	-1.90	-1.86	0.98
PHACTR4	-1.90	-1.84	0.97
TET2	-1.90	-2.14	1.13
KDM4C	-1.90	-2.23	1.17
AVL9	-1.90	-2.23	1.17
Hs,283851,0	-1.90	-3.53	1.85
BBX	-1.90	-2.36	1.24
CAB39L	-1.90	-3.27	1.72
RRAS2	-1.90	-1.72	0.90
GSK3B	-1.91	-1.76	0.93
PDLIM5	-1.91	-2.56	1.34
UBE2H	-1.91	-1.99	1.04
LOC147727	-1.91	-3.46	1.81
SNX9	-1.91	-2.77	1.45
PRKACB	-1.91	-2.10	1.10
SHOC2	-1.91	-1.79	0.94
GATAD2A	-1.91	-2.14	1.12
WDR25	-1.91	-2.46	1.29
PARN	-1.91	-2.15	1.12
DDX10	-1.91	-2.13	1.11
COCH	-1.92	-2.50	1.31
ATP8B1	-1.92	-2.35	1.22
CCNF	-1.92	-2.47	1.28
TYW1	-1.92	-2.28	1.19
CHSY1	-1.93	-1.67	0.87
PRAGMIN	-1.93	-1.83	0.95
EXOC6	-1.93	-2.54	1.32
FOXP1	-1.93	-1.79	0.93
CRTC3	-1.93	-2.09	1.08
CTBP2	-1.93	-2.19	1.13
LMBR1	-1.93	-3.41	1.77
ZNF572	-1.93	-3.10	1.61
UGCG	-1.93	-1.98	1.02
HECW2	-1.93	-3.13	1.62
TESK2	-1.93	-1.75	0.91

QKI	-1.93	-2.11	1.09
ARHGAP21	-1.93	-1.99	1.03
ZNF148	-1.94	-2.68	1.38
Hs,282093,1	-1.94	-1.58	0.82
ZNF322B	-1.94	-2.21	1.14
METT5D1	-1.94	-2.21	1.14
RYK	-1.94	-2.20	1.13
DAAM1	-1.94	-2.85	1.47
PSD3	-1.94	-3.06	1.58
BBX	-1.94	-2.46	1.27
UBR3	-1.94	-2.43	1.25
RTN4IP1	-1.94	-2.25	1.16
NR2C2	-1.94	-2.37	1.22
PPP2R2C	-1.94	-2.53	1.30
ITSN2	-1.94	-2.03	1.04
PHF2	-1.94	-2.16	1.11
TAF4B	-1.94	-2.39	1.23
GARNL1	-1.94	-2.40	1.23
IL4R	-1.95	-1.80	0.93
FBXO11	-1.95	-2.19	1.12
Hs,10450,1	-1.95	-2.41	1.23
TBL1X	-1.95	-2.38	1.22
PICALM	-1.95	-5.02	2.58
LYPD6B	-1.95	-2.11	1.08
Hs,18016,0	-1.95	-3.44	1.76
CTBP2	-1.95	-2.57	1.31
CAMSAP1L1	-1.95	-1.94	0.99
FLVCR2	-1.96	-2.27	1.16
NEAT1	-1.96	-2.38	1.22
DCDC2	-1.96	-2.36	1.20
UBXN7	-1.96	-2.23	1.14
LMBR1	-1.96	-2.19	1.12
GNAQ	-1.96	-2.79	1.42
SUFU	-1.96	-2.53	1.29
RSBN1L	-1.96	-3.21	1.63
STAG2	-1.97	-2.44	1.24
LARP4B	-1.97	-2.31	1.18

MAN1A1	-1.97	-2.52	1.28
TGFA	-1.97	-2.54	1.29
ANKRD50	-1.97	-2.25	1.14
NUAK1	-1.97	-2.15	1.09
PBRM1	-1.97	-2.54	1.29
ABL2	-1.98	-4.41	2.23
MEF2A	-1.98	-2.09	1.06
ANKRD13C	-1.98	-2.42	1.22
TRIM24	-1.98	-1.88	0.95
MTMR12	-1.98	-2.12	1.07
SIN3A	-1.98	-2.08	1.05
MKL2	-1.98	-2.23	1.12
FOXO1	-1.98	-2.96	1.49
EEA1	-1.99	-2.14	1.08
CTBP2	-1.99	-2.43	1.22
FAM20C	-1.99	-1.57	0.79
N4BP1	-1.99	-1.73	0.87
Hs,61426,1	-1.99	-2.19	1.10
PTPRK	-1.99	-2.45	1.23
TGFBR2	-1.99	-2.14	1.07
CUL1	-1.99	-2.27	1.14
WDR44	-1.99	-2.01	1.01
SDCCAG10	-1.99	-2.64	1.33
ZNRF1	-1.99	-2.15	1.08
ZCCHC2	-1.99	-2.24	1.13
KPNA3	-1.99	-2.36	1.18
GRHL3	-1.99	-1.66	0.83
GAB2	-1.99	-2.08	1.04
FUT8	-1.99	-2.36	1.18
PPP1R12A	-1.99	-2.18	1.09
PPP3CA	-1.99	-2.16	1.08
BMPR2	-2.00	-2.27	1.14
WWP1	-2.00	-2.06	1.03
PICALM	-2.00	-2.87	1.44
CRIM1	-2.00	-2.65	1.32
FOXK2	-2.00	-1.86	0.93
GRLF1	-2.00	-3.04	1.52

RREB1	-2.00	-1.88	0.94
SRPK2	-2.00	-2.69	1.34
C16orf52	-2.00	-2.07	1.03
ZHX3	-2.00	-2.54	1.27
PIK3R1	-2.01	-2.32	1.16
TNFSF10	-2.01	-2.98	1.49
Hs2,385784,1	-2.01	-2.15	1.07
GOLSYN	-2.01	-1.93	0.96
FCHO2	-2.01	-1.69	0.84
SPOPL	-2.01	-2.04	1.02
CDC42SE2	-2.01	-2.29	1.14
PIAS1	-2.01	-1.88	0.93
CSGALNACT1	-2.01	-2.08	1.03
ZC3H4	-2.01	-2.09	1.04
TBC1D9	-2.02	-2.23	1.11
GIGYF2	-2.02	-2.41	1.19
FAM53B	-2.02	-1.99	0.99
HS6ST2	-2.02	-2.57	1.28
CAMSAP1	-2.02	-2.31	1.15
DDHD1	-2.02	-2.31	1.14
tcag7,903	-2.02	-2.41	1.19
KIAA2026	-2.02	-2.12	1.05
PPAP2B	-2.02	-2.21	1.09
Hs2,253690,1	-2.02	-2.30	1.14
C11orf30	-2.02	-1.80	0.89
SESTD1	-2.02	-2.27	1.12
TIAM1	-2.02	-2.40	1.18
B4GALT5	-2.03	-1.73	0.86
TET3	-2.03	-2.22	1.10
TAF3	-2.03	-4.40	2.17
CENPA	-2.03	-2.48	1.22
CXXC5	-2.03	-2.39	1.18
CHD6	-2.03	-2.47	1.21
EPC2	-2.03	-2.07	1.02
ZNF680	-2.03	-2.10	1.03
CTBP2	-2.03	-2.51	1.24
LMBR1	-2.03	-3.11	1.53

CREBBP	-2.03	-2.05	1.01
MPRIP	-2.04	-2.42	1.19
ENOX2	-2.04	-2.39	1.18
SYNJ2	-2.04	-3.12	1.53
CBLB	-2.04	-2.30	1.13
HS6ST2	-2.04	-2.51	1.23
ORC2L	-2.04	-2.13	1.04
Hs,40334,0	-2.04	-1.88	0.92
NR3C1	-2.05	-2.25	1.10
GLCCI1	-2.05	-2.17	1.06
SPAG9	-2.05	-2.47	1.20
KIF1B	-2.05	-3.72	1.81
SLITRK6	-2.05	-1.98	0.97
NIPBL	-2.05	-2.46	1.20
LRIG1	-2.05	-2.90	1.41
TTC7A	-2.05	-1.98	0.96
GNA12	-2.06	-2.41	1.17
C1orf203	-2.06	-2.42	1.18
TACC2	-2.06	-2.39	1.16
GIGYF2	-2.06	-2.25	1.09
MAP4K4	-2.06	-2.26	1.09
MTUS1	-2.06	-2.46	1.19
DENND1B	-2.06	-3.75	1.82
ZCCHC14	-2.06	-2.48	1.20
STAG2	-2.07	-2.46	1.19
LYN	-2.07	-2.16	1.04
IRF2	-2.07	-2.28	1.10
PMEPA1	-2.07	-2.25	1.09
TEX2	-2.07	-2.39	1.15
Hs2,356537,1	-2.07	-4.59	2.21
NTN4	-2.07	-2.65	1.28
FARS2	-2.07	-3.14	1.51
MAP3K9	-2.07	-2.09	1.01
CABLES1	-2.08	-2.46	1.19
LOC730631	-2.08	-2.34	1.13
ASAP2	-2.08	-2.43	1.17
LRRC16A	-2.08	-3.20	1.54

SSBP2	-2.08	-3.25	1.56
LATS2	-2.08	-1.89	0.91
MAP4K4	-2.08	-3.36	1.61
RBMS1	-2.08	-2.44	1.17
Hs,168625,2	-2.08	-3.42	1.64
PDS5B	-2.09	-2.44	1.17
ATXN1	-2.09	-3.28	1.57
GAB1	-2.09	-2.92	1.40
CRIM1	-2.09	-3.77	1.80
FOXP1	-2.09	-3.84	1.84
CLASP2	-2.09	-2.41	1.15
CXXC5	-2.09	-2.52	1.20
NCOA1	-2.10	-2.98	1.42
SPIN1	-2.10	-2.19	1.04
LOH12CR1	-2.10	-2.70	1.29
SRPK2	-2.10	-2.48	1.18
RNF13	-2.10	-2.66	1.27
TLE1	-2.10	-2.33	1.11
LOC730631	-2.10	-2.72	1.29
HNRNPD	-2.10	-1.89	0.90
CEP350	-2.10	-2.17	1.03
ZC3H3	-2.10	-1.98	0.94
GALNT7	-2.10	-3.28	1.56
KITLG	-2.11	-2.08	0.99
ZHX2	-2.11	-2.54	1.20
PPP3CA	-2.11	-2.66	1.26
STXBP5	-2.11	-2.09	0.99
ERBB4	-2.11	-3.19	1.51
PLEKHF2	-2.11	-2.32	1.10
RSF1	-2.11	-2.41	1.14
KIAA1549	-2.11	-3.37	1.59
ZNF764	-2.12	-1.97	0.93
FOXO3	-2.12	-2.97	1.40
MPRIP	-2.12	-2.73	1.29
TMEM49	-2.12	-1.78	0.84
KLF7	-2.12	-1.83	0.87
LOC100127983	-2.12	-3.93	1.86

SIPA1L2	-2.12	-3.19	1.50
TTC39B	-2.12	-2.62	1.24
ZNF148	-2.12	-2.54	1.20
Hs,7589,0	-2.12	-2.63	1.24
METTL8	-2.12	-2.70	1.27
KIAA1432	-2.12	-2.26	1.06
MED13L	-2.13	-4.06	1.91
ARID2	-2.13	-2.84	1.33
PTPN3	-2.13	-2.84	1.33
HIVEP1	-2.14	-2.24	1.05
LGALS8	-2.14	-2.76	1.29
FOXN3	-2.14	-2.74	1.28
RBMS1	-2.14	-3.09	1.44
GCNT1	-2.14	-1.97	0.92
ATAD4	-2.14	-2.28	1.07
DIP2C	-2.14	-3.06	1.43
ANKRD11	-2.15	-2.71	1.26
ITGA2	-2.15	-2.99	1.39
GPD2	-2.15	-2.25	1.05
PCTK2	-2.15	-2.00	0.93
Hs,5985,1	-2.15	-3.26	1.52
SEC24B	-2.15	-2.12	0.99
TACC2	-2.15	-2.93	1.36
ARHGEF7	-2.16	-2.52	1.17
CCDC12	-2.16	-2.33	1.08
STAG1	-2.16	-2.24	1.04
Hs,171942,1	-2.16	-3.01	1.40
MEF2A	-2.16	-1.91	0.89
RAB28	-2.16	-2.35	1.09
SMARCA2	-2.16	-4.15	1.92
ARHGAP12	-2.16	-2.70	1.25
ZFYVE9	-2.16	-2.19	1.01
KIAA0922	-2.16	-2.42	1.12
Hs,322849,0	-2.16	-2.04	0.94
PIK3R3	-2.16	-2.01	0.93
CDC42SE2	-2.16	-2.77	1.28
SOCS5	-2.17	-1.94	0.90

RDX	-2.17	-2.56	1.18
RALGPS2	-2.17	-2.17	1.00
CTTNBP2NL	-2.17	-2.15	0.99
AKAP7	-2.17	-2.71	1.25
TAF4	-2.18	-2.43	1.12
PVRL3	-2.18	-2.53	1.16
C5orf41	-2.18	-2.74	1.26
RPS6KC1	-2.18	-2.48	1.14
MEGF9	-2.18	-3.03	1.39
SNX13	-2.19	-2.29	1.05
WWP1	-2.19	-2.51	1.15
EIF2C2	-2.19	-2.31	1.06
RCOR1	-2.19	-2.45	1.12
ATG5	-2.19	-3.32	1.51
FBXO34	-2.19	-1.91	0.87
C4orf19	-2.19	-3.28	1.50
CDC42SE2	-2.19	-3.02	1.37
Hs,233354,0	-2.20	-2.30	1.05
IFFO2	-2.20	-1.50	0.68
NCK1	-2.20	-2.43	1.10
MACC1	-2.20	-2.23	1.01
IGF1R	-2.21	-2.65	1.20
ATXN7	-2.21	-2.21	1.00
EHMT1	-2.21	-2.70	1.22
ATXN7L1	-2.21	-2.23	1.01
CTBP2	-2.21	-2.20	1.00
Hs,194110,1	-2.21	-2.20	1.00
UVRAG	-2.21	-2.36	1.07
SOCS5	-2.21	-1.92	0.87
ADCY1	-2.21	-4.30	1.94
BCAR3	-2.21	-2.36	1.06
ABCA1	-2.22	-2.07	0.93
ATRN	-2.22	-3.08	1.39
Hs,87134,0	-2.22	-2.14	0.97
RANBP9	-2.22	-2.73	1.23
MAP2K4	-2.22	-2.11	0.95
PHF15	-2.22	-2.43	1.09

142

LRP12	-2.22	-2.08	0.94
TBL1XR1	-2.23	-5.22	2.35
C5orf41	-2.23	-2.39	1.07
PTPN21	-2.23	-1.75	0.79
ATE1	-2.23	-2.24	1.01
ST8SIA4	-2.23	-2.53	1.14
ATRNL1	-2.23	-4.13	1.85
DTNBP1	-2.23	-2.29	1.02
SHROOM3	-2.24	-1.75	0.78
C10orf26	-2.24	-2.30	1.03
INSR	-2.24	-2.63	1.17
RYK	-2.24	-2.77	1.24
TBC1D30	-2.24	-1.83	0.82
POLR3B	-2.25	-2.56	1.14
GLI3	-2.25	-2.34	1.04
COL4A3BP	-2.25	-3.37	1.50
RPTOR	-2.25	-2.77	1.23
ARAP2	-2.25	-2.28	1.01
Hs,146275,0	-2.25	-1.54	0.68
Hs,184108,2	-2.25	-2.12	0.94
ARID2	-2.26	-2.91	1.29
CBFA2T3	-2.26	-1.97	0.87
SLITRK6	-2.26	-2.28	1.01
MAP4K3	-2.26	-2.56	1.14
GNAQ	-2.26	-2.88	1.28
FOXP1	-2.26	-2.03	0.90
DTNBP1	-2.26	-3.95	1.75
SRPK2	-2.26	-3.44	1.52
BANP	-2.26	-2.16	0.96
USP12	-2.26	-2.83	1.25
PDE5A	-2.26	-3.75	1.66
MEF2A	-2.26	-2.93	1.30
SKI	-2.26	-1.60	0.71
PTK2	-2.27	-2.61	1.15
CCNY	-2.27	-2.65	1.17
AMBRA1	-2.27	-2.43	1.07
NCEH1	-2.27	-2.46	1.08

BANP	-2.27	-2.41	1.06
EHBP1	-2.27	-2.57	1.13
ACVR1	-2.28	-2.41	1.06
SLCO3A1	-2.28	-2.54	1.11
GNAQ	-2.28	-2.93	1.29
PUM1	-2.28	-2.32	1.02
TANC2	-2.28	-3.23	1.41
IPPK	-2.28	-2.28	1.00
ARID4A	-2.28	-2.24	0.98
SPIRE1	-2.28	-2.55	1.12
UNQ1887	-2.28	-2.43	1.06
AMBRA1	-2.29	-2.83	1.24
SH3BP4	-2.29	-2.01	0.88
ZC3H10	-2.29	-2.53	1.10
C8orf44	-2.29	-5.93	2.59
SHROOM2	-2.29	-2.55	1.11
C1QTNF9	-2.30	-2.29	1.00
SFMBT1	-2.30	-2.44	1.06
ZNF764	-2.30	-2.23	0.97
ZBTB42	-2.31	-2.64	1.15
IRS1	-2.31	-2.60	1.13
STK3	-2.31	-4.55	1.97
ZNF708	-2.31	-2.72	1.18
SLITRK6	-2.31	-2.13	0.92
TIGD5	-2.31	-1.66	0.72
XPR1	-2.31	-2.80	1.21
TEAD1	-2.31	-3.34	1.44
ST3GAL1	-2.31	-3.52	1.52
NCK2	-2.31	-2.11	0.91
Hs,126889,0	-2.31	-1.97	0.85
PMEPA1	-2.31	-2.46	1.06
VAV3	-2.32	-3.30	1.42
MED26	-2.32	-2.29	0.99
SIPA1L2	-2.32	-4.30	1.85
Hs,18016,0	-2.33	-2.59	1.12
LOC220594	-2.33	-2.58	1.11
SESTD1	-2.33	-2.01	0.87

144

WDR70	-2.33	-2.34	1.00
SMURF1	-2.33	-2.24	0.96
RUNX2	-2.34	-3.07	1.31
UNQ1887	-2.34	-2.91	1.25
XRCC4	-2.34	-3.24	1.39
FIGN	-2.34	-3.07	1.31
ASAP1	-2.34	-3.16	1.35
ZFP36L2	-2.34	-2.83	1.21
ARHGEF10L	-2.34	-2.15	0.92
ASXL2	-2.35	-2.53	1.08
SASH1	-2.35	-3.02	1.29
HIPK2	-2.35	-2.86	1.22
FBXW11	-2.35	-2.11	0.90
MBP	-2.35	-2.58	1.10
TBC1D8	-2.36	-1.97	0.84
RBMS1	-2.36	-3.11	1.32
CASZ1	-2.36	-3.02	1.28
ARID4B	-2.36	-2.25	0.95
FAM73A	-2.36	-2.32	0.98
TUBD1	-2.36	-4.30	1.82
FAM73A	-2.37	-4.22	1.79
RAPGEF1	-2.37	-3.36	1.42
MYST3	-2.37	-2.37	1.00
DIP2C	-2.37	-2.55	1.08
PIK3CB	-2.37	-2.72	1.15
PIP4K2A	-2.37	-2.52	1.07
RBM47	-2.37	-3.90	1.64
RASSF8	-2.37	-3.13	1.32
LRRC1	-2.38	-2.60	1.09
ZMIZ1	-2.38	-2.51	1.05
SAP30	-2.38	-10.66	4.48
FUT9	-2.38	-2.64	1.11
TRPS1	-2.38	-3.26	1.37
RAPGEF6	-2.39	-2.59	1.09
PLEKHA5	-2.40	-2.90	1.21
PSPC1	-2.40	-2.40	1.00
DSCAM	-2.40	-2.51	1.05

AGAP1	-2.40	-2.69	1.12
MAP3K5	-2.40	-4.58	1.90
TECPR2	-2.40	-3.17	1.32
KIAA0240	-2.41	-2.75	1.15
ARHGAP17	-2.41	-2.30	0.95
NEDD4L	-2.41	-3.56	1.48
TASP1	-2.41	-2.64	1.09
SPIRE1	-2.41	-3.71	1.54
FAM134B	-2.41	-2.82	1.17
FAM102A	-2.41	-3.62	1.50
Hs,77273,2	-2.41	-3.57	1.48
DCBLD1	-2.41	-2.12	0.88
RNF144B	-2.41	-2.01	0.83
KIAA1267	-2.42	-2.30	0.95
ABHD11	-2.42	-3.19	1.32
MLLT10	-2.42	-3.14	1.29
NR3C1	-2.44	-2.25	0.93
SH2D4A	-2.44	-2.56	1.05
NR1H3	-2.45	-2.09	0.86
NLK	-2.45	-2.78	1.14
FUT8	-2.45	-4.70	1.92
CLIP1	-2.45	-4.60	1.88
APBB2	-2.45	-3.29	1.35
RSBN1L	-2.45	-2.45	1.00
DYRK1A	-2.45	-2.40	0.98
RSF1	-2.45	-2.51	1.02
ASCL1	-2.45	-3.83	1.56
RFX7	-2.45	-2.37	0.97
BMP7	-2.46	-2.44	0.99
PTK2	-2.46	-5.33	2.17
MGAT5	-2.46	-5.42	2.20
BANP	-2.47	-2.82	1.14
MLLT3	-2.47	-2.34	0.95
CSNK1G1	-2.47	-2.92	1.18
SH3RF1	-2.47	-2.57	1.04
LDLR	-2.47	-3.57	1.44
MED13L	-2.47	-2.86	1.16

SGMS1	-2.47	-2.88	1.16
SSH2	-2.48	-2.90	1.17
B4GALT5	-2.48	-2.31	0.93
USP32	-2.48	-2.76	1.12
LDLRAD3	-2.48	-3.56	1.44
EPHA4	-2.48	-2.53	1.02
NEDD9	-2.48	-3.39	1.37
KIAA0247	-2.48	-2.31	0.93
TBC1D12	-2.48	-2.68	1.08
PPM1H	-2.49	-3.23	1.30
POU2F1	-2.50	-2.74	1.10
VANGL1	-2.50	-3.14	1.26
TLE4	-2.50	-2.62	1.05
RNGTT	-2.50	-3.35	1.34
TGFB2	-2.51	-2.43	0.97
RBMS1	-2.51	-4.00	1.59
RAPGEF6	-2.51	-2.63	1.05
Hs,213547,0	-2.51	-2.50	0.99
EXT1	-2.51	-3.07	1.22
SPTLC2	-2.51	-5.22	2.08
Hs,158196,2	-2.52	-2.82	1.12
MBNL1	-2.52	-3.67	1.46
TNFRSF11B	-2.52	-2.91	1.16
Hs,29131,1	-2.52	-2.92	1.16
EVL	-2.52	-3.46	1.38
ZNF407	-2.52	-3.10	1.23
FGFR2	-2.52	-3.05	1.21
FAM59A	-2.53	-3.61	1.43
GLCCI1	-2.53	-3.75	1.48
FAM126B	-2.53	-2.74	1.08
GPATCH2	-2.54	-2.72	1.07
TBC1D30	-2.54	-2.45	0.96
ATG10	-2.54	-3.15	1.24
MSL2	-2.55	-2.74	1.08
DSCAM	-2.55	-3.63	1.42
MFHAS1	-2.55	-2.36	0.93
VAV3	-2.55	-2.78	1.09

PIAS1	-2.56	-2.76	1.08
BAZ2B	-2.56	-2.61	1.02
ABL2	-2.57	-3.68	1.43
FOXJ3	-2.57	-2.11	0.82
MEIS2	-2.57	-2.86	1.11
TLE1	-2.57	-4.71	1.83
PDS5B	-2.58	-3.38	1.31
LYPD6	-2.58	-2.83	1.09
Hs,59368,0	-2.59	-2.37	0.92
EYA2	-2.59	-2.36	0.91
DNAJC6	-2.59	-3.21	1.24
CDYL	-2.60	-2.89	1.11
DSCAM	-2.60	-3.17	1.22
RPS6KA5	-2.61	-3.66	1.40
XRCC4	-2.61	-3.99	1.53
ST8SIA4	-2.61	-2.79	1.07
SPRED2	-2.61	-3.09	1.18
HOMER1	-2.61	-3.34	1.28
RPS6KA5	-2.62	-2.71	1.04
SPEN	-2.62	-3.25	1.24
KLF7	-2.62	-2.74	1.04
APBB2	-2.62	-3.27	1.25
ZNF678	-2.62	-3.08	1.18
KIAA0240	-2.62	-3.79	1.44
ADCY9	-2.63	-3.20	1.22
NRIP1	-2.64	-3.52	1.33
ATL2	-2.64	-2.72	1.03
SULF1	-2.64	-3.46	1.31
ORC5L	-2.65	-3.68	1.39
KDM6A	-2.66	-3.43	1.29
ZMYND8	-2.66	-3.37	1.27
GRHL2	-2.66	-3.25	1.22
TMTC2	-2.66	-3.19	1.20
GRLF1	-2.66	-2.82	1.06
ZFP36L2	-2.67	-3.41	1.28
ZSWIM6	-2.68	-3.42	1.28
TMEM51	-2.68	-2.32	0.87

148

MLL3	-2.69	-3.93	1.46
GNAQ	-2.69	-2.69	1.00
Hs,21914,0	-2.69	-2.78	1.03
PHF20	-2.70	-3.06	1.13
LDLR	-2.71	-3.60	1.33
VGLL4	-2.72	-2.52	0.92
ZFHX3	-2.73	-2.98	1.09
RXRA	-2.73	-3.26	1.19
MAGI3	-2.73	-4.13	1.51
ANKS1A	-2.74	-3.39	1.24
RUNX1	-2.75	-4.47	1.63
LRP6	-2.75	-3.38	1.23
C3orf59	-2.76	-3.21	1.16
THRB	-2.77	-2.83	1.02
MED27	-2.77	-3.17	1.15
Hs,176376,0	-2.78	-4.15	1.49
TLK2	-2.78	-2.51	0.90
KALRN	-2.78	-2.84	1.02
LARGE	-2.78	-3.51	1.26
PPARA	-2.79	-2.01	0.72
MEIS1	-2.79	-3.15	1.13
PTPRJ	-2.80	-3.87	1.38
TPST1	-2.80	-3.93	1.41
MYST4	-2.80	-2.72	0.97
EFNA5	-2.80	-3.75	1.34
LOC729436	-2.80	-2.40	0.86
NFATC1	-2.81	-2.58	0.92
CHD7	-2.81	-2.72	0.97
Hs,112157,0	-2.81	-3.61	1.29
TLK2	-2.81	-4.03	1.43
JMJD1C	-2.81	-2.98	1.06
LYN	-2.81	-2.66	0.95
ATXN1	-2.82	-4.25	1.51
NRP1	-2.82	-3.60	1.28
Hs,197617,0	-2.83	-4.30	1.52
RASAL2	-2.84	-3.71	1.31
FNIP1	-2.84	-2.45	0.86

INO80D	-2.84	-3.03	1.07
XRCC4	-2.85	-5.58	1.96
CDYL	-2.87	-5.02	1.75
PUM1	-2.88	-3.99	1.38
FOXN3	-2.89	-3.89	1.35
JARID2	-2.90	-3.45	1.19
APBB2	-2.91	-3.08	1.06
PRKCH	-2.92	-4.02	1.38
NRIP1	-2.92	-3.48	1.19
LIN52	-2.92	-2.30	0.79
ASAP1	-2.93	-3.13	1.07
NHSL1	-2.94	-3.86	1.32
C7orf60	-2.94	-3.29	1.12
SYNCRIP	-2.95	-2.74	0.93
DIP2B	-2.96	-3.75	1.27
TRIO	-2.96	-3.13	1.06
RHPN2	-2.98	-2.47	0.83
MLLT3:	-2.98	-2.92	0.98
RAB28	-2.99	-9.45	3.16
NCOA3	-2.99	-4.62	1.55
MAST4	-3.00	-3.78	1.26
IRS1	-3.00	-3.20	1.07
ST3GAL1	-3.00	-3.15	1.05
SASH1	-3.01	-3.47	1.16
LCORL	-3.01	-3.36	1.11
SRGAP1	-3.01	-4.60	1.53
SRPK2	-3.01	-4.50	1.49
RAI1	-3.04	-2.80	0.92
GSK3B	-3.04	-4.03	1.33
PSD3	-3.04	-3.97	1.30
SYT1	-3.04	-4.40	1.45
INPP5A	-3.06	-4.02	1.32
RBMS1	-3.06	-3.35	1.10
KIAA0355	-3.07	-2.71	0.88
MAST4	-3.07	-3.75	1.22
CDK6	-3.07	-3.38	1.10
GNA14	-3.08	-5.69	1.85

MAP3K5	-3.09	-2.88	0.93
RNGTT	-3.09	-4.47	1.44
USP6NL	-3.10	-3.55	1.14
TMCC1	-3.11	-3.19	1.03
Hs,102558,0	-3.12	-3.85	1.23
CDK6	-3.12	-3.70	1.19
Hs,236894,0	-3.14	-5.13	1.63
RASGRP1	-3.14	-4.83	1.54
Hs,57079,0	-3.17	-6.47	2.04
Hs,118769,0	-3.18	-4.31	1.36
RAPGEF2	-3.18	-3.41	1.07
TNFAIP8	-3.18	-3.29	1.03
Hs,5415,0	-3.20	-2.51	0.78
RBM47	-3.21	-4.95	1.54
TNFAIP8	-3.21	-3.11	0.97
BMPR1A	-3.25	-4.30	1.32
TMEM117	-3.26	-3.87	1.19
EFNA5	-3.34	-5.04	1.51
FNIP2	-3.34	-4.02	1.20
Hs,152423,0	-3.34	-4.05	1.21
CHST15	-3.35	-3.06	0.92
LHFPL2	-3.35	-4.30	1.28
SLC20A2	-3.38	-6.27	1.86
RSF1	-3.39	-3.48	1.03
HIVEP2 protein 2	-3.39	-3.49	1.03
Hs,94499,0	-3.40	-3.55	1.04
NCOA3	-3.42	-6.64	1.94
TMTC2	-3.46	-7.80	2.25
TMCC1	-3.46	-4.95	1.43
IKZF2	-3.47	-4.40	1.27
HS3ST3A1	-3.48	-4.10	1.18
SHB	-3.51	-3.69	1.05
MBD5	-3.52	-5.00	1.42
FOXO1	-3.54	-3.42	0.97
TNFRSF11B	-3.56	-5.93	1.67
STK3	-3.57	-4.12	1.15
KCNJ3	-3.59	-4.26	1.19

TRPS1	-3.60	-5.53	1.54
Hs,199776,0	-3.60	-3.83	1.06
Hs,99283,0	-3.61	-3.84	1.06
BMPR1A	-3.62	-3.70	1.02
ST8SIA4	-3.63	-3.78	1.04
SEMA4D	-3.64	-4.32	1.19
LYN	-3.65	-6.74	1.84
Hs,19215,0	-3.67	-4.96	1.35
SRGAP1	-3.68	-10.74	2.92
KLF12	-3.70	-5.00	1.35
TCF7L2	-3.74	-5.97	1.60
NRF1	-3.76	-3.47	0.92
FNDC3B	-3.79	-5.43	1.43
C4orf19	-3.80	-3.94	1.04
NRF1	-3.84	-5.14	1.34
SCAPER	-3.87	-5.18	1.34
EPHA4	-3.90	-5.19	1.33
LRCH1	-3.90	-4.16	1.07
SRGAP1	-3.93	-3.66	0.93
ARHGEF3	-3.94	-5.89	1.50
TMCC1	-3.95	-4.50	1.14
SYT1	-3.98	-7.02	1.76
TANC1	-3.98	-5.09	1.28
PCDH7	-4.04	-6.80	1.68
STARD13	-4.07	-6.36	1.56
THRB	-4.19	-5.36	1.28
MMP16	-4.20	-4.37	1.04
FGFR2	-4.29	-7.84	1.83
PDZRN3	-4.30	-5.28	1.23
JARID2	-4.31	-4.37	1.01
TAF3	-4.34	-4.26	0.98
ARID5B	-4.34	-4.16	0.96
TET3	-4.41	-3.82	0.86
C1orf226	-4.43	-6.24	1.41
Hs,235857,0	-4.48	-4.42	0.99
Hs,323099,0	-4.58	-3.61	0.79
HS6ST3	-4.59	-5.50	1.20

FCHSD2	-4.64	-3.75	0.81
GLI3	-4.68	-5.04	1.08
TCF7L2	-4.70	-4.40	0.93
Hs,60257,0	-4.76	-4.90	1.03
TCF7L2	-4.86	-4.80	0.99
TNS3	-4.89	-5.49	1.12
FGD4	-5.07	-5.69	1.12
TCF7L2	-5.08	-7.08	1.39
RIN2	-5.16	-9.02	1.75
ARHGAP24	-5.61	-7.34	1.31
MAML2	-5.71	-7.26	1.27
GPR39	-7.44	-12.38	1.66
ZNRF3	-9.35	-7.14	0.76
FAM125B	-13.85	-16.85	1.22

Printed in the United States
By Bookmasters